NFPA POCKET GUIDE TO FIRE ALARM SYSTEM INSTALLATION

MERTON W. BUNKER, JR.
RICHARD J. ROUX

NFPA
Quincy, Massachusetts

Product Manager: Brad Gray
Editorial-Production Services: Omegatype Typography, Inc.
Interior Design/Composition: Omegatype Typography, Inc.
Cover Design: Lana Kurtz Graphic Design
Manufacturing Manager: Ellen Glisker
Printer: Rose Printing Company

Copyright © 2002
National Fire Protection Association, Inc.
One Batterymarch Park
Quincy, Massachusetts 02269

Notice Concerning Liability: Publication of this work is for the purpose of circulating information and opinion among those concerned for
fire safety and related subjects. While every effort has been made to
achieve a work of high quality, neither the NFPA nor the authors and
contributors to this work guarantee the accuracy or completeness of or
assume any liability in connection with the information and opinions
contained in this work. The NFPA and the authors and contributors
shall in no event be liable for any personal injury, property, or other
damages of any nature whatsoever, whether special, indirect, conse-
quential, or compensatory, directly or indirectly resulting from the publication, use of or reliance upon this work.

This work is published with the understanding that the NFPA and the
authors and contributors to this work are supplying information and
opinion but are not attempting to render engineering or other professional services. If such services are required, the assistance of an
appropriate professional should be sought.

NFPA No.: PGFAS-01
ISBN: 0-87765-521-9
Library of Congress Control Number: 2002101406

Printed in the United States of America
06 05 04 03 02   5 4 3

Credits:

Exhibits II.8 (p. 53), II.9 (p. 54), II.10 (p. 55): Reprinted from Donald G. Fink and H. Wayne Beaty, eds., *Standard Handbook for Electrical Engineers,* 12th ed., by permission of McGraw-Hill.

Exhibit III.50 (pp. 148–151): Reprinted from 2001 ASHRAE Handbook-Fundamentals, by permission of American Society of Heating, Refrigerating and Air-Conditioning Engineers, Inc.

Exhibits VIII.6 (p. 542), VIII.7 (p. 543), VIII.8 (pp. 544–545), VIII.9 (p. 546), VIII.10 (p. 547), VIII.11 (p. 548), VIII.12 (p. 549), VIII.13 (p. 550), VIII.14 (p. 551), VIII.15 (pp. 552–554), VIII.16 (p. 555), VIII.17 (pp. 556–559), VIII.18 (pp. 560–561): Reprinted from NECA 100-1999, *Symbols for Electrical Construction Drawings* (ANSI) by permission of National Electrical Contractors Association, Bethesda, MD.

Fire Alarm–Related Definitions ❚ Emergency Systems Definitions

System Types ❚ Power Supplies ❚ Circuits ❚ Alarm
Sequence ❚ Control Functions ❚ Signals ❚ Survivability

Detection Checklist ❚ Ceiling Surfaces ❚ Heat Detector Temperature
Classification ❚ Heat Detector Mounting and Spacing ❚ Smoke
Detector Response Characteristics ❚ Smoke Detector Mounting and
Spacing ❚ Smoke Detectors in High Air Movement Areas ❚ Smoke
Detectors for Smoke Control

Notification Appliance Mounting ❚ Audible Notification
Appliances ❚ Visible Notification Appliances ❚ Visible Equivalency

Performance Criteria ❚ Communications Methods

Test Checklist ❚ Methods and Frequencies ❚ Battery Voltage

Field Connections ❚ Electrical Relationships ❚ Conductors ❚
Cables ❚ Conduit and Raceway Systems

BRIEF CONTENTS

BRIEF CONTENTS

CONTENTS

CONTENTS   xiii

HEAT DETECTOR TEMPERATURE CLASSIFICATION 103

HEAT DETECTOR MOUNTING AND SPACING 104

SMOKE DETECTORS IN HIGH AIR MOVEMENT AREAS   134

SMOKE DETECTORS FOR SMOKE CONTROL   136

SECTION IV   155
NOTIFICATION APPLIANCES

NOTIFICATION APPLIANCE MOUNTING   156

AUDIBLE NOTIFICATION APPLIANCES 157

VISIBLE NOTIFICATION APPLIANCES 165

SECTION V   181
SUPERVISING STATIONS

SECTION VII
WIRING
259

CONTENTS

SECTION VIII
DOCUMENTATION AND DRAWINGS   509

SYSTEM DOCUMENTATION   510

ELECTRICAL DRAWING SYMBOLS   541

INTRODUCTION

The information in this *NFPA Pocket Guide to Fire Alarm System Installation* has been compiled to provide a portable, handy reference source for the field practitioner. Intended for engineers, designers, contractors, installers, inspectors, loss control specialists, and students, the *Pocket Guide* offers valuable information for the design, review, evaluation, installation, testing, and maintenance of a fire alarm or other fire protection system.

Fires have cost many American lives, and tremendous property losses have been encountered and documented since the country was colonized in the seventeenth century. Early detection of fires was, and still is, critical in saving lives and protecting property. Detection is also necessary to begin suppression activities.

Before the advent of automatic fire detection and notification, sentinels, known as "watchmen," roamed the streets of cities and towns or staffed watchtowers during the night. If a fire was spotted, they alerted others by ringing church bells. Fire watches also used horns, drums, and even wooden rattles to alert sleeping fire fighters. These practices were the beginning of fire alarm signaling.

The first public fire alarm reporting system, then known as the Boston Fire Alarm Telegraph, was placed in service at noon on April 28, 1852. Dr. William F. Channing and Moses G. Farmer developed this system in 1845, just one year after Samuel Morse invented the telegraph. When a fire was spotted by a sentinel, the alarm was raised by operating a hand crank signal wheel, which in turn transmitted the box number to the central station, actuating the system. The first fire alarm was reported at 8:25 P.M. on this system the very next day.

This telegraph system had 40 cast iron street boxes and 19 tower bell strikers and was used for nearly 16 years before being replaced. The system was developed utilizing telegraph techniques because telephones were not widely used in the city of Boston until about 1874. More modern cousins of the telegraph system are still in use today, some 150 years later.

The first automatic fire alarm detection devices included fusible link type (nonrestorable) heat detectors. These were first developed around 1849. Notification appliances received their start with the introduction of the alarm bell, also developed by Moses Farmer, in 1848.

The National Fire Protection Association was founded in 1896, in response to the need for sprinkler and electrical wiring installation standards, and standards for other technologies used in industry. Poor practices by installers of sprinkler systems and electrical wiring resulted in large losses for the manufacturing sector at that time. The lack of standardization led to the creation of NFPA 13, *Standard for the Installation of Sprinkler Systems*. In an effort to expand standardization of installation practices to other areas of fire protection, NFPA developed the first signaling standard in 1899. NFPA 71-D, *General Rules and Requirements for the Installation of Wiring and Apparatus for Automatic Fire Alarms, Hatch Closers, Sprinkler Alarms, and Other Automatic Alarm Systems and Their Manual Auxiliaries,* sought to create uniformity for all fire alarm signaling system installations. Since that time, fire alarms systems have gained the reputation of a valuable and effective tool for fire protection.

Although incremental improvements were made to many existing technologies, fire detection and notification systems changed very little from the beginning of the twentieth century until the 1960s. It was at that time that smoke detectors first became listed and used in commercial applications. The use of smoke detectors spread to dwellings in the 1970s and since then has accounted for saving thousands of lives annually. Other forms of detection include radiant energy–sensing fire detectors, gas detectors, and air sampling type smoke detectors. Other technology advancements include visible notification appliances, widely used since the early 1980s.

Modern fire alarm systems are microprocessor- and software-controlled marvels of technology. They can sense the smallest of fire conditions and alert occupants and fire brigades while the fire is still at its incipient stages. In addition, they monitor and control other systems such as elevators, sprinkler systems, chemical and water extinguishing systems, and fire pumps. Modern fire alarm systems can also control other building systems such as heating, ventilating, and air conditioning (HVAC) systems, to make the premises safer during a fire emergency. Unlike their early predecessors, they do this with a minimal amount of human intervention and effort.

With the advancement of technology, requirements for fire alarm installations as well as the available materials and equipment have become increasingly complex. The material included in this *Pocket Guide* was chosen based on the problems and situations frequently encountered in the field, where it is not practical to transport reference material. However, the *Pocket Guide* is not intended to be used as a substitute for the actual applicable codes and standards, nor is it a training manual for untrained persons. It is intended to provide a portable, handy reference for the field practitioner. *NFPA 72®*, *National Fire Alarm Code®*, contains many additional requirements not included in the *Pocket Guide*. The following is a brief summary of the *Pocket Guide*'s ten sections.

Section I, "Definitions," contains selected definitions from NFPA 70, *National Electrical Code®*, *NFPA 72®*, *National Fire Alarm Code®*, and NFPA *101®*, *Life Safety Code®*.

Section II, "Fundamentals," offers information relating to major common system requirements, including system types, power supply requirements, battery calculations, circuit class and style, monitoring for integrity, circuits extending beyond one building, alarm sequence, control functions, signals, and survivability.

Section III, "Initiating Devices," provides various requirements for the spacing, location, and installation of commonly used initiating devices. Initiating devices covered include heat detectors, line-type heat detectors, and smoke detectors. This information is frequently needed in the field for verification of proper device placement.

Section IV, "Notification Appliances," provides basic data to assist the user in design, spacing, location, and installation of both audible and visible notification appliances. Included are tables and figures showing sound pressure levels, useful relationships, light criteria, and visible equivalency. This information, which is commonly used by design and installation personnel, is useful throughout the design and installation process.

Section V, "Supervising Stations," is a collection of basic requirements for supervising station systems and related communications methods. Included are figures and tables showing the criteria for several supervising station types, transmission methods, loading, and other criteria.

Section VI, "Inspection, Testing, and Maintenance," includes test checklists and tables showing test methods, testing and visual inspection frequencies, and voltage criteria for nickel cadmium and lead acid batteries. This section contains tables and schedules as well as basic information necessary for the performance of overall fire alarm system testing. This information is needed to establish verification of test methods in the field.

Section VII, "Wiring," provides essential data for the designer and installer relevant to wiring methods, conductors, cable substitutions, and conduit and raceway systems. This section contains voltage drop and loop resistance information; conductor and cable data; conduit and box fill requirements; and specifications for power-limited, Class 2, and Class 3 power supplies. For explanation, correct and incorrect wiring methods are illustrated and compared. A primer on commonly used electrical relationships is included. Basic technical criteria, relationships, and formulas are presented for inductance, capacitance, and impedance.

Section VIII, "Documentation and Drawings," provides reference material and useful information on working plans for fire alarm systems. Included is a collection of commonly used symbols for electrical, fire alarm, and fire protection devices from various sources, including NFPA 170, *Standard for Fire Safety Symbols*, and NECA 100-1999, *Symbols for Electrical Construction Drawings*. It identifies the information needed on drawings and illustrates commonly used symbols for devices, fire alarm control units and appliances. This information is critical to anyone who must interpret the notations of fire alarm system elements on system documentation. This information is included in an effort to encourage the use of standardized symbols on fire alarm drawings.

Section IX, "Useful Tables, Formulas, and Figures," provides the user with conversion factors between SI and U.S. customary units of measurement. This information is especially important since most NFPA codes and standards are being revised to use SI units. Common conversion factors for length, area, volume, light, and temperature are included. General engineering formulas that are useful to anyone in the field who needs to initiate calculations or verify estimates are provided. Basic data are provided to assist in identifying device and appliance characteristics. This information is needed by anyone associated with the design, installation, or commissioning of a fire alarm system.

Section X, "Useful Contacts," is a listing of organizations and contact information useful to designers, installers, and authorities having jurisdiction.

NFPA POCKET GUIDE TO FIRE ALARM SYSTEM INSTALLATION

SECTION I

DEFINITIONS

This section contains selected definitions from NFPA 70, *National Electrical Code*®, NFPA *72*®, *National Fire Alarm Code*®, and NFPA *101*®, *Life Safety Code*®. It provides a convenient reference for contractors and deals exclusively with the terms for installing system components such as fire alarm control units; initiating devices; notification appliances; monitor and control of other systems such as elevators, sprinkler systems, chemical and water extinguishing systems, and fire pumps; and interfaces to other building systems, such as heating, ventilating, and air conditioning (HVAC) systems.

FIRE ALARM–RELATED DEFINITIONS

Editor's Note: The definitions in this section have been extracted from the following documents: NFPA 70, National Electrical Code®, 2002 edition; NFPA 72®, National Fire Alarm Code®, 1999 edition; and NFPA 101®, Life Safety Code®, 2000 edition.

Accessible (as applied to equipment). Admitting close approach; not guarded by locked doors, elevation, or other effective means. [70]

Accessible (as applied to wiring methods). Capable of being removed or exposed without damaging the building structure or finish or not permanently closed in by the structure or finish of the building. [70]

Accessible, Readily (Readily Accessible). Capable of being reached quickly for operation, renewal, or inspections, without requiring those to whom ready access is requisite to climb over or remove obstacles or to resort to portable ladders, and so forth. [70]

Acknowledge. To confirm that a message or signal has been received, such as by the pressing of a button or the selection of a software command. [72]

Active Multiplex System. A multiplexing system in which signaling devices such as transponders are employed to transmit status signals of each initiating device or initiating device circuit within a prescribed time interval so that the lack of receipt of such a signal can be interpreted as a trouble signal. [72]

Addressable Device. A fire alarm system component with discrete identification that can have its status individually identified or that is used to individually control other functions. [72]

Adverse Condition. Any condition occurring in a communications or transmission channel that interferes with the proper transmission or interpretation, or both, of status change signals at the supervising station. *(Refer to Trouble Signal.)* [72]

Air Sampling–Type Detector. A detector that consists of a piping or tubing distribution network that runs from the detector to the area(s) to be protected. An aspiration fan in the detector housing draws air from the protected area back to the detector through air sampling ports, piping, or tubing. At the detector, the air is analyzed for fire products. [72]

Alarm. A warning of fire danger. [72]

Alarm Service. The service required following the receipt of an alarm signal. [72]

Alarm Signal. A signal indicating an emergency that requires immediate action, such as a signal indicative of fire. [72]

Alarm Verification Feature. A feature of automatic fire detection and alarm systems to reduce unwanted alarms wherein smoke detectors report alarm conditions for a minimum period of time, or confirm alarm conditions within a given time period after being reset, in order to be accepted as a valid alarm initiation signal. [72]

Alert Tone. An attention-getting signal to alert occupants of the pending transmission of a voice message. [72]

Analog Initiating Device (Sensor). An initiating device that transmits a signal indicating varying degrees of condition as contrasted with a conventional initiating device, which can only indicate an on–off condition. [72]

Annunciator. A unit containing one or more indicator lamps, alphanumeric displays, or other equivalent means in which each indication provides status information about a circuit, condition, or location. [72]

Area, Gross Leasable. The total floor area designated for tenant occupancy and exclusive use, expressed in square feet (square meters), measured from the centerlines of adjoining partitions and exteriors of outside walls. [101]

Area, Hazardous. An area of a structure or building that poses a degree of hazard greater than that normal to the

FIRE ALARM—RELATED DEFINITIONS

general occupancy of the building or structure, such as areas used for the storage or use of combustibles or flammables; toxic, noxious, or corrosive materials; or heat-producing appliances. [101]

Area, Living. Any normally occupiable space in a residential occupancy, other than sleeping rooms or rooms that are intended for combination sleeping/living, bathrooms, toilet compartments, kitchens, closets, halls, storage or utility spaces, and similar areas. [101]

Area of Refuge. An area that is either (1) a story in a building where the building is protected throughout by an approved, supervised automatic sprinkler system and has not less than two accessible rooms or spaces separated from each other by smoke-resisting partitions; or (2) a space located in a path of travel leading to a public way that is protected from the effects of fire, either by means of separation from other spaces in the same building or by virtue of location, thereby permitting a delay in egress travel from any level. [101]

Area of Refuge, Accessible. An area of refuge that complies with the accessible route requirements of CABO/ANSI A117.1, *American National Standard for Accessible and Usable Buildings and Facilities.* [101]

Atrium. A large-volume space created by a floor opening or series of floor openings connecting two or more stories that is covered at the top of the series of openings and is used for purposes other than an enclosed stairway; elevator hoistway; escalator opening; or utility shaft used for plumbing, electrical, air-conditioning, or communications facilities. [101]

Audible Notification Appliance. A notification appliance that alerts by the sense of hearing. [72]

Automatic Extinguishing System Supervisory Device. A device that responds to abnormal conditions that could affect the proper operation of an automatic sprinkler system or other fire extinguishing system(s) or suppression system(s), including, but not limited to, control valves; pressure levels; liquid agent levels and temperatures; pump power and running; engine temperature and overspeed; and room temperature. [72]

Automatic Fire Detector. A device designed to detect the presence of a fire signature and to initiate action. For the purpose of this code, automatic fire detectors are classified as follows: Automatic Fire Extinguishing or Suppression System Operation Detector, Fire-Gas Detector, Heat Detector, Other Fire Detectors, Radiant Energy-Sensing Fire Detector, Smoke Detector. [72]

Automatic Fire Extinguishing or Suppression System Operation Detector. A device that automatically detects the operation of a fire extinguishing or suppression system by means appropriate to the system employed. [72]

Auxiliary Box. A fire alarm box that can be operated from one or more remote actuating devices. [72]

Auxiliary Fire Alarm System. A system connected to a municipal fire alarm system for transmitting an alarm of fire to the public fire service communications center. Fire alarms from an auxiliary fire alarm system are received at the public fire service communications center on the same equipment and by the same methods as alarms transmitted manually from municipal fire alarm boxes located on streets. [72]

Auxiliary Fire Alarm System, Local Energy Type. An auxiliary system that employs a locally complete arrangement of parts, initiating devices, relays, power supply, and associated components to automatically trip a municipal transmitter or master box over electrical circuits that are electrically isolated from the municipal system circuits. [72]

Auxiliary Fire Alarm System, Parallel Telephone Type. An auxiliary system connected by a municipally controlled individual circuit to the protected property to interconnect the initiating devices at the protected premises and the municipal fire alarm switchboard. [72]

FIRE ALARM–RELATED DEFINITIONS

Auxiliary Fire Alarm System, Shunt Auxiliary Type. An auxiliary system electrically connected to an integral part of the municipal alarm system extending the municipal circuit into the protected premises to interconnect the initiating devices, which, when operated, open the municipal circuit shunted around the trip coil of the municipal transmitter or master box. The municipal transmitter or master box is thereupon energized to start transmission without any assistance from a local source of power. **[72]**

Average Ambient Sound Level. The root mean square, A-weighted, sound pressure level measured over a 24-hour period. **[72]**

Barrier, Smoke. A continuous membrane, or a membrane with discontinuities created by protected openings, where such membrane is designed and constructed to restrict the movement of smoke. **[101]**

Barrier, Thermal. A material that limits the average temperature rise of an unexposed surface to not more than 250°F (139°C) for a specified fire exposure complying with the standard time-temperature curve of NFPA 251, *Standard Methods of Tests of Fire Endurance of Building Construction and Materials*. **[101]**

Box Battery. The battery supplying power for an individual fire alarm box where radio signals are used for the transmission of box alarms. **[72]**

Branch Circuit. The circuit conductors between the final overcurrent device protecting the circuit and the outlet(s). **[70]**

Building, Apartment. A building containing three or more dwelling units with independent cooking and bathroom facilities. **[101]**

Building, Bulk Merchandising Retail. A building in which the sales area includes the storage of combustible materials on pallets, in solid piles, or in racks in excess of 12 ft (3.7 m) in storage height. **[101]**

Building, Covered Mall. A building, including the covered mall, enclosing a number of tenants and occupancies wherein two or more tenants have a main entrance into the covered mall. **[101]**

Building, Existing. A building erected or officially authorized prior to the effective date of the adoption of this edition of the *Code* by the agency or jurisdiction. **[101]**

Building, High-Rise. A building greater than 75 ft (23 m) in height where the building height is measured from the lowest level of fire department vehicle access to the floor of the highest occupiable story. **[101]**

Building, Historic. A structure and its associated additions and site deemed to have historical, architectural, or cultural significance by a local, regional, or national jurisdiction. **[101]**

Building, Special Amusement. A building that is temporary, permanent, or mobile that contains a device or system that conveys passengers or provides a walkway along, around, or over a course in any direction as a form of amusement arranged so that the egress path is not readily apparent due to visual or audio distractions or an intentionally confounded egress path, or is not readily available due to the mode of conveyance through the building or structure. **[101]**

Carrier. High-frequency energy that can be modulated by voice or signaling impulses. **[72]**

Carrier System. A means of conveying a number of channels over a single path by modulating each channel on a different carrier frequency and demodulating at the receiving point to restore the signals to their original form. **[72]**

Ceiling. The upper surface of a space, regardless of height. Areas with a suspended ceiling have two ceilings, one visible from the floor and one above the suspended ceiling. **[72]**

Ceiling Height. The height from the continuous floor of a room to the continuous ceiling of a room or space. **[72]**

FIRE ALARM–RELATED DEFINITIONS

Ceiling Surfaces, Beam Construction. Ceilings that have solid structural or solid nonstructural members projecting down from the ceiling surface more than 4 in. (100 mm) and spaced more than 3 ft (0.9 m), center to center. [72]

Ceiling Surfaces, Girder. A support for beams or joists that runs at right angles to the beams or joists. If the top of the girder is within 4 in. (100 mm) of the ceiling, the girder is a factor in determining the number of detectors and is to be considered a beam. If the top of the girder is more than 4 in. (100 mm) from the ceiling, the girder is not a factor in detector location. [72]

Ceiling Surfaces, Solid Joist Construction. Ceilings that have solid structural or solid nonstructural members projecting down from the ceiling surface for a distance of more than 4 in. (100 mm) and spaced at intervals of 3 ft (0.9 m) or less, center to center. [72]

Central Station. A supervising station that is listed for central station service. [72]

Central Station Fire Alarm System. A system or group of systems in which the operations of circuits and devices are transmitted automatically to, recorded in, maintained by, and supervised from a listed central station that has competent and experienced servers and operators who, upon receipt of a signal, take such action as required by this code. Such service is to be controlled and operated by a person, firm, or corporation whose business is the furnishing, maintaining, or monitoring of supervised fire alarm systems. [72]

Central Station Service. The use of a system or a group of systems in which the operations of circuits and devices at a protected property are signaled to, recorded in, and supervised from a listed central station that has competent and experienced operators who, upon receipt of a signal, take such action as required by this code. Related activities at the protected property, such as equipment installation, inspection, testing, maintenance, and runner service, are the responsibility of the central station or a listed fire alarm service–local company. Central station service is controlled and operated by a person, firm, or corporation whose business is the furnishing of such contracted services or whose properties are the protected premises. [72]

Certification. A systematic program that uses randomly selected follow-up inspections of the certified systems installed under the program that allows the listing organization to verify that a fire alarm system complies with all the requirements of this code. A system installed under such a program is identified by the issuance of a certificate and is designated as a certificated system. [72]

Certification of Personnel. A formal program of related instruction and testing as provided by a recognized organization or the authority having jurisdiction. [72]

Channel. A path for voice or signal transmission that uses modulation of light or alternating current within a frequency band. [72]

Circuit Interface. A circuit component that interfaces initiating devices or control circuits, or both; notification appliances or circuits, or both; system control outputs; and other signaling line circuits to a signaling line circuit. [72]

Cloud Chamber Smoke Detection. The principle of using an air sample drawn from the protected area into a high-humidity chamber combined with a lowering of chamber pressure to create an environment in which the resultant moisture in the air condenses on any smoke particles present, forming a cloud. The cloud density is measured by a photoelectric principle. The density signal is processed and used to convey an alarm condition when it meets preset criteria. [72]

Coded. An audible or visible signal that conveys several discrete bits or units of information. Notification signal examples are numbered strokes of an impact-type appliance and numbered flashes of a visible appliance. [72]

Combination Detector. A device that either responds to more than one of the fire phenomenon or employs more

than one operating principle to sense one of these phenomenon. Typical examples are a combination of a heat
detector with a smoke detector or a combination rate-of-
rise and fixed-temperature heat detector. [72]

Combination Fire Alarm and Guard's Tour Box. A manually operated box for separately transmitting a fire alarm signal and a distinctive guard patrol tour supervisory signal.
[72]

Combination System. A fire alarm system in which components are used, in whole or in part, in common with a
non-fire signaling system. [72]

Combustible. A material that, in the form in which it is
used and under the conditions anticipated, will ignite and
burn; a material that does not meet the definition of noncombustible or limited-combustible. [101]

Combustion. A chemical process that involves oxidation
sufficient to produce light or heat. [101]

Communications Channel. A circuit or path connecting a
subsidiary station(s) to a supervising station(s) over which
signals are carried. [72]

Compatibility Listed. A specific listing process that ap-
plies only to two-wire devices, such as smoke detectors,
that are designed to operate with certain control equipment. [72]

Compatible Equipment. Equipment that interfaces mechanically or electrically as manufactured without field
modification. [72]

Contiguous Property. A single-owner or single-user protected premises on a continuous plot of ground, including
any buildings thereon, that is not separated by a public
thoroughfare, transportation right-of-way, property owned
or used by others, or body of water not under the same
ownership. [72]

Control Unit. A system component that monitors inputs
and controls outputs through various types of circuits. [72]

Court. An open, uncovered, unoccupied space, unobstructed to the sky, bounded on three or more sides by exterior building walls. [101]

Court, Enclosed. A court bounded on all sides by the exterior walls of a building or by the exterior walls and lot lines on which walls are permitted. [101]

Covered Mall. A covered or roofed interior area used as a pedestrian way and connected to a building(s) or portions of a building housing single or multiple tenants. [101]

Day-Care Home. A building or portion of a building in which more than three but not more than 12 clients receive care, maintenance, and supervision, by other than their relative(s) or legal guardians(s), for less than 24 hours per day. [101]

Delinquency Signal. A signal indicating the need for action in connection with the supervision of guards or system attendants. [72]

Derived Channel. A signaling line circuit that uses the local leg of the public switched network as an active multiplex channel while simultaneously allowing that leg's use for normal telephone communications. [72]

Detector. A device suitable for connection to a circuit that has a sensor that responds to a physical stimulus such as heat or smoke. [72]

Digital Alarm Communicator Receiver (DACR). A system component that accepts and displays signals from digital alarm communicator transmitters (DACTs) sent over the public switched telephone network. [72]

Digital Alarm Communicator System (DACS). A system in which signals are transmitted from a digital alarm communicator transmitter (DACT) located at the protected premises through the public switched telephone network to a digital alarm communicator receiver (DACR). [72]

Digital Alarm Communicator Transmitter (DACT). A system component at the protected premises to which initiating devices or groups of devices are connected. The

FIRE ALARM–RELATED DEFINITIONS

DACT seizes the connected telephone line, dials a prese-lected number to connect to a DACR, and transmits signals indicating a status change of the initiating device. **[72]**

Digital Alarm Radio Receiver (DARR). A system compo-nent composed of two subcomponents: one that receives and decodes radio signals, the other that annunciates the decoded data. These two subcomponents can be coresi-dent at the central station or separated by means of a data transmission channel. **[72]**

Digital Alarm Radio System (DARS). A system in which signals are transmitted from a digital alarm radio transmit-ter (DART) located at a protected premises through a radio channel to a digital alarm radio receiver (DARR). **[72]**

Digital Alarm Radio Transmitter (DART). A system com-ponent that is connected to or an integral part of a digital alarm communicator transmitter (DACT) that is used to provide an alternate radio transmission channel. **[72]**

Display. The visual representation of output data, other than printed copy. **[72]**

Dormitory. A building or a space in a building in which group sleeping accommodations are provided for more than 16 persons who are not members of the same family in one room or a series of closely associated rooms under joint occupancy and single management, with or without meals, but without individual cooking facilities. **[101]**

Double Doorway. A single opening that has no interven-ing wall space or door trim separating the two doors. **[72]**

Double Dwelling Unit. A building consisting solely of two dwelling units. *(See Dwelling Unit.)* **[72]**

Draft Stop. A continuous membrane used to subdivide a concealed space to restrict the passage of smoke, heat, and flames. **[101]**

Dual Control. The use of two primary trunk facilities over separate routes or different methods to control one com-munications channel. **[72]**

FIRE ALARM–RELATED DEFINITIONS

Dwelling Unit. One or more rooms for the permanent use of one or more persons as a space for eating, living, and sleeping, with permanent provisions for cooking and sanitation. For the purposes of this code, *dwelling unit* includes one- and two-family attached and detached dwellings, apartments, and condominiums but does not include hotel and motel rooms and guest suites, dormitories, or sleeping rooms in nursing homes. **[72]**

Dwelling Unit. A single unit, providing complete, independent living facilities for one or more persons, including permanent provisions for living, sleeping, eating, cooking, and sanitation. **[101]**

Electrical Conductivity Heat Detector. A line-type or spot-type sensing element in which resistance varies as a function of temperature. **[72]**

Elevator Lobby. A space from which people directly enter an elevator car(s) and to which people directly leave an elevator car(s). **[101]**

Elevator Lobby Door. A door between an elevator lobby and another building space other than the elevator shaft. **[101]**

Ember. A particle of solid material that emits radiant energy due either to its temperature or the process of combustion on its surface. *(Refer to Spark.)* **[72]**

Emergency Access Opening. A window, panel, or similar opening in which (1) the opening has dimensions of not less than 22 in. (55.9 cm) in width and 24 in. (61 cm) in height and is unobstructed to allow for ventilation and rescue operations from the exterior, (2) the bottom of the opening is not more than 44 in. (112 cm) above the floor, (3) the opening is readily identifiable from both the exterior and interior, and (4) the opening is readily openable from both the exterior and interior. **[101]**

Emergency Voice/Alarm Communications. Dedicated manual or automatic facilities for originating and distributing voice instructions, as well as alert and evacuation signals pertaining to a fire emergency, to the occupants of a building. [72]

Evacuation. The withdrawal of occupants from a building. [72]

Evacuation Signal. Distinctive signal intended to be recognized by the occupants as requiring evacuation of the building. [72]

Exit. That portion of a means of egress that is separated from all other spaces of a building or structure by construction or equipment as required to provide a protected way of travel to the exit discharge. [101]

Exit Access. That portion of a means of egress that leads to an exit. [101]

Exit Discharge. That portion of a means of egress between the termination of an exit and a public way. [101]

Exit Discharge, Level of. (1) The lowest story from which not less than 50 percent of the required number of exits and not less than 50 percent of the required egress capacity from such a story discharge directly outside at grade; (2) the story with the smallest elevation change needed to reach grade where no story has 50 percent or more of the required number of exits and 50 percent or more of the required egress capacity from such a story discharge directly outside at grade. [101]

Exit, Horizontal. A way of passage from one building to an area of refuge in another building on approximately the same level, or a way of passage through or around a fire barrier to an area of refuge on approximately the same level in the same building that affords safety from fire and smoke originating from the area of incidence and areas communicating therewith. [101]

Exit Plan. A plan for the emergency evacuation of the premises. [72]

Field of View. The solid cone that extends out from the detector within which the effective sensitivity of the detector is at least 50 percent of its on-axis, listed, or approved sensitivity. **[72]**

Fire Alarm Circuit. The portion of the wiring system between the load side of the overcurrent device or the power-limited supply and the connected equipment of all circuits powered and controlled by the fire alarm system. Fire alarm circuits are classified as either non–power-limited or power-limited. **[70]**

Fire Alarm Circuit Integrity (CI) Cable. Cable used in fire alarm systems to ensure continued operation of critical circuits during a specified time under fire conditions. **[70]**

Fire Alarm Control Unit (Panel). A system component that receives inputs from automatic and manual fire alarm devices and might supply power to detection devices and to a transponder(s) or off-premises transmitter(s). The control unit might also provide transfer of power to the notification appliances and transfer of condition to relays or devices connected to the control unit. The fire alarm control unit can be a local fire alarm control unit or a master control unit. **[72]**

Fire Alarm/Evacuation Signal Tone Generator. A device that produces a fire alarm/evacuation tone upon command. **[72]**

Fire Alarm Signal. A signal initiated by a fire alarm-initiating device such as a manual fire alarm box, automatic fire detector, waterflow switch, or other device in which activation is indicative of the presence of a fire or fire signature. **[72]**

Fire Alarm System. A system or portion of a combination system that consists of components and circuits arranged to monitor and annunciate the status of fire alarm or supervisory signal-initiating devices and to initiate the appropriate response to those signals. **[72]**

Fire Barrier. A continuous membrane or a membrane with discontinuities created by protected openings with a specified fire protection rating, where such membrane is designed and constructed with a specified fire resistance rating to limit the spread of fire and that also restricts the movement of smoke. [101]

Fire Command Center. The principal attended or unattended location where the status of the detection, alarm communications, and control systems is displayed and from which the system(s) can be manually controlled. [72]

Fire-Gas Detector. A device that detects gases produced by a fire. [72]

Fire Rating. The classification indicating in time (hours) the ability of a structure or component to withstand a standardized fire test. This classification does not necessarily reflect performance of rated components in an actual fire. [72]

Fire Safety Function Control Device. The fire alarm system component that directly interfaces with the control system that controls the fire safety function. [72]

Fire Safety Functions. Building and fire control functions that are intended to increase the level of life safety for occupants or to control the spread of the harmful effects of fire. [72]

Fire Warden. A building staff member or a tenant trained to perform assigned duties in the event of a fire emergency. [72]

Fixed-Temperature Detector. A device that responds when its operating element becomes heated to a predetermined level. [72]

Flame. A body or stream of gaseous material involved in the combustion process and emitting radiant energy at specific wavelength bands determined by the combustion chemistry of the fuel. In most cases, some portion of the emitted radiant energy is visible to the human eye. [72]

Flame Detector. A radiant energy-sensing fire detector that detects the radiant energy emitted by a flame. [72]

FIRE ALARM–RELATED DEFINITIONS

Flame Detector Sensitivity. The distance along the optical axis of the detector at which the detector can detect a fire of specified size and fuel within a given time frame. **[72]**

Flashover. A stage in the development of a contained fire in which all exposed surfaces reach ignition temperatures more or less simultaneously and fire spreads rapidly throughout the space. **[101]**

Fuel Load. The total quantity of combustible contents of a building, space, or fire area. **[101]**

Gateway. A device that is used in the transmission of serial data (digital or analog) from the fire alarm control unit to other building system control units, equipment, or networks and/or from other building system control units to the fire alarm control unit. **[72]**

Goal. A nonspecific overall outcome to be achieved that is measured on a qualitative basis. **[101]**

Guard's Tour Reporting Station. A device that is manually or automatically initiated to indicate the route being followed and the timing of a guard's tour. **[72]**

Guard's Tour Supervisory Signal. A supervisory signal monitoring the performance of guard patrols. **[72]**

Guest Room. An accommodation combining living, sleeping, sanitary, and storage facilities within a compartment. **[101]**

Guest Suite. An accommodation with two or more contiguous rooms comprising a compartment, with or without doors between such rooms, that provides living, sleeping, sanitary, and storage facilities. **[101]**

Heat Alarm. A single or multiple station alarm responsive to heat. **[72]**

Heat Detector. A fire detector that detects either abnormally high temperature or rate of temperature rise, or both. **[72]**

Hoistway. Any shaftway, hatchway, well hole, or other vertical opening or space in which an elevator or dumbwaiter is designed to operate. **[70]**

Hospital. A building or portion thereof used on a 24-hour basis for the medical, psychiatric, obstetrical, or surgical care of four or more inpatients. **[101]**

Hotel. A building or groups of buildings under the same management in which there are sleeping accommodations for more than 16 persons and primarily used by transients for lodging with or without meals. **[101]**

Household Fire Alarm System. A system of devices that produces an alarm signal in the household for the purpose of notifying the occupants of the presence of a fire so that they will evacuate the premises. **[72]**

Hunt Group. A group of associated telephone lines within which an incoming call is automatically routed to an idle (not busy) telephone line for completion. **[72]**

Initiating Device. A system component that originates transmission of a change-of-state condition, such as in a smoke detector, manual fire alarm box, or supervisory switch. **[72]**

Initiating Device Circuit. A circuit to which automatic or manual initiating devices are connected where the signal received does not identify the individual device operated. **[72]**

Intermediate Fire Alarm or Fire Supervisory Control Unit. A control unit used to provide area fire alarm or area fire supervisory service that, where connected to the proprietary fire alarm system, becomes a part of that system. **[72]**

Ionization Smoke Detection. The principle of using a small amount of radioactive material to ionize the air between two differentially charged electrodes to sense the presence of smoke particles. Smoke particles entering the ionization volume decrease the conductance of the air by reducing ion mobility. The reduced conductance signal is processed and used to convey an alarm condition when it meets preset criteria. **[72]**

Labeled. Equipment or materials to which has been attached a label, symbol, or other identifying mark of an organization that is acceptable to the authority having jurisdiction and concerned with product evaluation, that maintains periodic inspection of production of labeled equipment or materials, and by whose labeling the manufacturer indicates compliance with appropriate standards or performance in a specified manner. **[72]**

Leg Facility. The portion of a communications channel that connects not more than one protected premises to a primary or secondary trunk facility. The leg facility includes the portion of the signal transmission circuit from its point of connection with a trunk facility to the point where it is terminated within the protected premises at one or more transponders. **[72]**

Level Ceilings. Ceilings that are level or have a slope of less than or equal to 1 in 8. **[72]**

Life Safety Network. A type of combination system that transmits fire safety control data through gateways to other building system control units. **[72]**

Limited Care Facility. A building or portion of a building used on a 24-hour basis for the housing of four or more persons who are incapable of self-preservation because of age; physical limitations due to accident or illness; or limitations such as mental retardation/developmental disability, mental illness, or chemical dependency. **[101]**

Limited-Combustible. Refers to a building construction material not complying with the definition of *noncombustible* that, in the form in which it is used, has a potential heat value not exceeding 3500 Btu/lb (8141 kJ/kg), where tested in accordance with NFPA 259, *Standard Test Method for Potential Heat of Building Materials*, and includes (1) materials having a structural base of noncombustible material, with a surfacing not exceeding a

FIRE ALARM–RELATED DEFINITIONS

thickness of 1/8 in. (3.2 mm) that has a flame spread index not greater than 50; and (2) materials, in the form and thickness used, other than as described in (1), having neither a flame spread index greater than 25 nor evidence of continued progressive combustion, and of such composition that surfaces that would be exposed by cutting through the material on any plane would have neither a flame spread index greater than 25 nor evidence of continued progressive combustion. [101]

Line-Type Detector. A device in which detection is continuous along a path. Typical examples are rate-of-rise pneumatic tubing detectors, projected beam smoke detectors, and heat-sensitive cable. [72]

Listed. Equipment, materials, or services included in a list published by an organization that is acceptable to the authority having jurisdiction and concerned with evaluation of products or services, that maintains periodic inspection of production of listed equipment or materials or periodic evaluation of services, and whose listing states that either the equipment, material, or service meets appropriate designated standards or has been tested and found suitable for a specified purpose. [72]

Loading Capacity. The maximum number of discrete elements of fire alarm systems permitted to be used in a particular configuration. [72]

Location, Damp. Locations protected from weather and not subject to saturation with water or other liquids but subject to moderate degrees of moisture. Examples of such locations include partially protected locations under canopies, marquees, roofed open porches, and like locations, and interior locations subject to moderate degrees of moisture, such as some basements, some barns, and some cold storage warehouses. [70]

Location, Dry. A location not normally subject to dampness or wetness. A location classified as dry may be temporarily subject to dampness or wetness, as in the case of a building under construction. [70]

FIRE ALARM–RELATED DEFINITIONS

Location, Wet. Installations under ground or in concrete slabs or masonry in direct contact with the earth; in locations subject to saturation with water or other liquids, such as vehicle washing areas; and in unprotected locations exposed to weather. **[70]**

Lodging or Rooming House. A building or portion thereof that does not qualify as a one- or two-family dwelling, that provides sleeping accommodations for a total of 16 or fewer people on a transient or permanent basis, without personal care services, with or without meals, but without separate cooking facilities for individual occupants. **[101]**

Loss of Power. The reduction of available voltage at the load below the point at which equipment can function as designed. **[72]**

Low-Power Radio Transmitter. Any device that communicates with associated control/receiving equipment by low-power radio signals. **[72]**

Maintenance. Repair service, including periodic inspections and tests, required to keep the fire alarm system and its component parts in an operative condition at all times, and the replacement of the system or its components when they become undependable or inoperable for any reason. **[72]**

Manual Fire Alarm Box. A manually operated device used to initiate an alarm signal. **[72]**

Master Box. A municipal fire alarm box that can also be operated by remote means. **[72]**

Master Control Unit (Panel). A control unit that serves the protected premises or portion of the protected premises as a local control unit and accepts inputs from other fire alarm control units. **[72]**

Means of Egress. A continuous and unobstructed way of travel from any point in a building or structure to a public way consisting of three separate and distinct parts: (1) the exit access, (2) the exit, and (3) the exit discharge. **[101]**

FIRE ALARM–RELATED DEFINITIONS

Means of Egress, Accessible. A path of travel, usable by a person with a severe mobility impairment, that leads to a public way or an area of refuge. **[101]**

Multiple Dwelling Unit. A building containing three or more dwelling units. *(See Dwelling Unit.)* **[72]**

Multiple Station Alarm. A single station alarm capable of being interconnected to one or more additional alarms so that the actuation of one causes the appropriate alarm signal to operate in all interconnected alarms. **[72]**

Multiple Station Alarm Device. Two or more single station alarm devices that can be interconnected so that actuation of one causes all integral or separate audible alarms to operate; or one single station alarm device having connections to other detectors or to a manual fire alarm box. **[72]**

Multiplexing. A signaling method characterized by simultaneous or sequential transmission, or both, and reception of multiple signals on a signaling line circuit, a transmission channel, or a communications channel, including means for positively identifying each signal. **[72]**

Municipal Fire Alarm Box (Street Box). An enclosure housing a manually operated transmitter used to send an alarm to the public fire service communications center. **[72]**

Municipal Fire Alarm System. A system of alarm-initiating devices, receiving equipment, and connecting circuits (other than a public telephone network) used to transmit alarms from street locations to the public fire service communications center. **[72]**

Municipal Transmitter. A transmitter that can only be tripped remotely that is used to send an alarm to the public fire service communications center. **[72]**

Noncoded Signal. An audible or visible signal conveying one discrete bit of information. **[72]**

Noncombustible. Refers to a material that, in the form in which it is used and under the conditions anticipated, does

FIRE ALARM–RELATED DEFINITIONS

not ignite, burn, support combustion, or release flammable vapors, when subjected to fire or heat. Materials that are reported as passing ASTM E 136, *Standard Test Method for Behavior of Materials in a Vertical Tube Furnace at 750 Degrees C*, are considered noncombustible materials. [101]

Noncontiguous Property. An owner- or user-protected premises where two or more protected premises, controlled by the same owner or user, are separated by a public thoroughfare, body of water, transportation right-of-way, or property owned or used by others. [72]

Nonrequired System. A supplementary fire alarm system component or group of components that is installed at the option of the owner, and is not installed due to a building or fire code requirement. [72]

Nonrestorable Initiating Device. A device in which the sensing element is designed to be destroyed in the process of operation. [72]

Notification Appliance. A fire alarm system component such as a bell, horn, speaker, light, or text display that provides audible, tactile, or visible outputs, or any combination thereof. [72]

Notification Appliance Circuit. A circuit or path directly connected to a notification appliance(s). [72]

Notification Zone. An area covered by notification appliances that are activated simultaneously. [72]

Nuisance Alarm. Any alarm caused by mechanical failure, malfunction, improper installation, or lack of proper maintenance, or any alarm activated by a cause that cannot be determined. [72]

Nursing Home. A building or portion of a building used on a 24-hour basis for the housing and nursing care of four or more persons who, because of mental or physical incapacity, might be unable to provide for their own needs and safety without the assistance of another person. [101]

Occupancy. The purpose for which a building or portion thereof is used or intended to be used. **[101]**

Occupancy, Ambulatory Health Care. A building or portion thereof used to provide services or treatment simultaneously to four or more patients that (1) provides, on an outpatient basis, treatment for patients that renders the patients incapable of taking action for self-preservation under emergency conditions without the assistance of others; or (2) provides, on an outpatient basis, anesthesia that renders the patients incapable of taking action for self-preservation under emergency conditions without the assistance of others. **[101]**

Occupancy, Assembly. An occupancy (1) used for a gathering of 50 or more persons for deliberation, worship, entertainment, eating, drinking, amusement, awaiting transportation, or similar uses; or (2) used as a special amusement building, regardless of occupant load. **[101]**

Occupancy, Business. An occupancy used for account and record keeping or the transaction of business other than mercantile. **[101]**

Occupancy, Day-Care. An occupancy in which four or more clients receive care, maintenance, and supervision, by other than their relatives or legal guardians, for less than 24 hours per day. **[101]**

Occupancy, Detention and Correctional. An occupancy used to house four or more persons under varied degrees of restraint or security where such occupants are mostly incapable of self-preservation because of security measures not under the occupants' control. **[101]**

Occupancy, Educational. An occupancy used for educational purposes through the twelfth grade by six or more persons for four or more hours per day or more than 12 hours per week. **[101]**

Occupancy, Health Care. An occupancy used for purposes of medical or other treatment or care of four or more persons where such occupants are mostly incapable of self-preservation due to age, physical or mental disability, or because of security measures not under the occupants' control. [101]

Occupancy, Industrial. An occupancy in which products are manufactured or in which processing, assembling, mixing, packaging, finishing, decorating, or repair operations are conducted. [101]

Occupancy, Industrial, General. An industrial occupancy in which ordinary and low hazard industrial operations are conducted in buildings of conventional design suitable for various types of industrial processes. [101]

Occupancy, Industrial, High Hazard. An industrial occupancy in which industrial operations that include high hazard materials, processes, or contents are conducted. [101]

Occupancy, Industrial, Special Purpose. An industrial occupancy in which ordinary and low hazard industrial operations are conducted in buildings designed for and suitable only for particular types of operations, characterized by a relatively low density of employee population, with much of the area occupied by machinery or equipment. [101]

Occupancy, Mercantile. An occupancy used for the display and sale of merchandise. [101]

Occupancy, Mixed. An occupancy in which two or more classes of occupancy exist in the same building or structure and where such classes are intermingled so that separate safeguards are impracticable. [101]

Occupancy, Multipurpose Assembly. An assembly room designed to accommodate temporarily any of several possible assembly uses. [101]

Occupancy, Residential. An occupancy that provides sleeping accommodations for purposes other than health care or detention and correctional. [101]

Occupancy, Residential Board and Care. A building or portion thereof that is used for lodging and boarding of four or more residents, not related by blood or marriage to the owners or operators, for the purpose of providing personal care services. [101]

Occupancy, Storage. An occupancy used primarily for the storage or sheltering of goods, merchandise, products, vehicles, or animals. [101]

Off-Hook. To make connection with the public-switched telephone network in preparation for dialing a telephone number. [72]

On-Hook. To disconnect from the public-switched telephone network. [72]

Open Area Detection (Protection). Protection of an area such as a room or space with detectors to provide early warning of fire. [72]

Operating Mode, Private. Audible or visible signaling only to those persons directly concerned with the implementation and direction of emergency action initiation and procedure in the area protected by the fire alarm system. [72]

Operating Mode, Public. Audible or visible signaling to occupants or inhabitants of the area protected by the fire alarm system. [72]

Operating System Software. The basic operating system software that can be altered only by the equipment manufacturer or its authorized representative. Operating system software is sometimes referred to as *firmware, BIOS,* or *executive program.* [72]

Other Fire Detectors. Devices that detect a phenomenon other than heat, smoke, flame, or gases produced by a fire. [72]

Ownership. Any property or building or its contents under legal control by the occupant, by contract, or by holding of a title or deed. [72]

Paging System. A system intended to page one or more persons by such means as voice over loudspeaker, coded audible signals or visible signals, or lamp annunciators. [72]

Parallel Telephone System. A telephone system in which an individually wired circuit is used for each fire alarm box. [72]

Path (Pathways). Any conductor, optic fiber, radio carrier, or other means for transmitting fire alarm system information between two or more locations. [72]

Permanent Visual Record (Recording). An immediately readable, not easily alterable, print, slash, or punch record of all occurrences of status change. [72]

Photoelectric Light Obscuration Smoke Detection. The principle of using a light source and a photosensitive sensor onto which the principal portion of the source emissions is focused. When smoke particles enter the light path, some of the light is scattered and some is absorbed, thereby reducing the light reaching the receiving sensor. The light reduction signal is processed and used to convey an alarm condition when it meets preset criteria. [72]

Photoelectric Light-Scattering Smoke Detection. The principle of using a light source and a photosensitive sensor arranged so that the rays from the light source do not normally fall onto the photosensitive sensor. When smoke particles enter the light path, some of the light is scattered by reflection and refraction onto the sensor. The light signal is processed and used to convey an alarm condition when it meets preset criteria. [72]

Placarded. A means to signify that the fire alarm system of a particular facility is receiving central station service in accordance with this code by a listed central station or listed fire alarm service—local company that is part of a systematic follow-up program under the control of an independent third party listing organization. [72]

Plant. One or more buildings under the same ownership or control on a single property. [72]

FIRE ALARM–RELATED DEFINITIONS

Plenum. A compartment or chamber to which one or more air ducts are connected and that forms part of the air distribution system. [101]

Pneumatic Rate-of-Rise Tubing Heat Detector. A line-type detector comprising small-diameter tubing, usually copper, that is installed on the ceiling or high on the walls throughout the protected area. The tubing is terminated in a detector unit containing diaphragms and associated contacts set to actuate at a predetermined pressure. The system is sealed except for calibrated vents that compensate for normal changes in temperature. [72]

Positive Alarm Sequence. An automatic sequence that results in an alarm signal, even when manually delayed for investigation, unless the system is reset. [72]

Power Supply. A source of electrical operating power, including the circuits and terminations connecting it to the dependent system components. [72]

Primary Battery (Dry Cell). A nonrechargeable battery requiring periodic replacement. [72]

Primary Trunk Facility. That part of a transmission channel connecting all leg facilities to a supervising or subsidiary station. [72]

Prime Contractor. The one company contractually responsible for providing central station services to a subscriber as required by this code. The prime contractor can be either a listed central station or a listed fire alarm service–local company. [72]

Private Radio Signaling. A radio system under control of the proprietary supervising station. [72]

Projected Beam–Type Detector. A type of photoelectric light obscuration smoke detector wherein the beam spans the protected area. [72]

Proprietary Supervising Station. A location to which alarm or supervisory signaling devices on proprietary fire alarm systems are connected and where personnel are in attendance at all times to supervise operation and investigate signals. [72]

Proprietary Supervising Station Fire Alarm System. An installation of fire alarm systems that serves contiguous and noncontiguous properties, under one ownership, from a proprietary supervising station located at the protected property, at which trained, competent personnel are in constant attendance. This includes the proprietary supervising station; power supplies; signal-initiating devices; initiating device circuits; signal notification appliances; equipment for the automatic, permanent visual recording of signals; and equipment for initiating the operation of emergency building control services. [72]

Protected Premises. The physical location protected by a fire alarm system. [72]

Protected Premises (Local) Control Unit (Panel). A control unit that serves the protected premises or a portion of the protected premises and indicates the alarm via notification appliances inside the protected premises. [72]

Protected Premises (Local) Fire Alarm System. A protected premises system that sounds an alarm at the protected premises as the result of the manual operation of a fire alarm box or the operation of protection equipment or systems, such as water flowing in a sprinkler system, the discharge of carbon dioxide, the detection of smoke, or the detection of heat. [72]

Public Fire Alarm Reporting System. A system of fire alarm initiating devices, receiving equipment, and connecting circuits used to transmit alarms from street locations to the communications center. [72]

Public Fire Alarm Reporting System, Type A. A system in which an alarm from a fire alarm box is received and is retransmitted to fire stations either manually or automatically. [72]

Public Fire Alarm Reporting System, Type B. A system in which an alarm from a fire alarm box is automatically transmitted to fire stations and, if used, is transmitted to supplementary alerting devices. [72]

Public Fire Service Communications Center. The building or portion of the building used to house the central operating part of the fire alarm system; usually the place where the necessary testing, switching, receiving, transmitting, and power supply devices are located. [72]

Public Switched Telephone Network. An assembly of communications facilities and central office equipment operated jointly by authorized common carriers that provides the general public with the ability to establish communications channels via discrete dialing codes. [72]

Radiant Energy–Sensing Fire Detector. A device that detects radiant energy (such as ultraviolet, visible, or infrared) that is emitted as a product of combustion reaction and obeys the laws of optics. [72]

Radio Alarm Repeater Station Receiver (RARSR). A system component that receives radio signals and resides at a repeater station that is located at a remote receiving location. [72]

Radio Alarm Supervising Station Receiver (RASSR). A system component that receives data and annunciates that data at the supervising station. [72]

Radio Alarm System (RAS). A system in which signals are transmitted from a radio alarm transmitter (RAT) located at a protected premises through a radio channel to two or more radio alarm repeater station receivers (RARSR) and that are annunciated by a radio alarm supervising station receiver (RASSR) located at the central station. [72]

Radio Alarm Transmitter (RAT). A system component at the protected premises to which initiating devices or groups of devices are connected that transmits signals indicating a status change of the initiating devices. [72]

Radio Channel. A band of frequencies of a width sufficient to allow its use for radio communications. [72]

Rate Compensation Detector. A device that responds when the temperature of the air surrounding the device reaches a predetermined level, regardless of the rate of temperature rise. **[72]**

Rate-of-Rise Detector. A device that responds when the temperature rises at a rate exceeding a predetermined value. **[72]**

Rating, Fire Protection. The designation indicating the duration of the fire test exposure to which a fire door assembly or fire window assembly was exposed and for which it met all the acceptance criteria as determined in accordance with NFPA 252, *Standard Methods of Fire Tests of Door Assemblies*, or NFPA 257, *Standard on Fire Test for Window and Glass Block Assemblies*, respectively. **[101]**

Rating, Fire Resistance. The time, in minutes or hours, that materials or assemblies have withstood a fire exposure as established in accordance with the test procedures of NFPA 251, *Standard Methods of Tests of Fire Endurance of Building Construction and Materials*. **[101]**

Record Drawings. Drawings (as-built) that document the location of all devices, appliances, wiring sequences, wiring methods, and connections of the components of the fire alarm system as installed. **[72]**

Record of Completion. A document that acknowledges the features of installation, operation (performance), service, and equipment with representation by the property owner, system installer, system supplier, service organization, and the authority having jurisdiction. **[72]**

Relocation. The movement of occupants from a fire zone to a safe area within the same building. **[72]**

Remote Supervising Station Fire Alarm System. A system installed in accordance with this code to transmit alarm, supervisory, and trouble signals from one or more protected premises to a remote location where appropriate action is taken. **[72]**

FIRE ALARM–RELATED DEFINITIONS

Repeater Station. The location of the equipment needed to relay signals between supervising stations, subsidiary stations, and protected premises. **[72]**

Reset. A control function that attempts to return a system or device to its normal, non-alarm state. **[72]**

Restorable Initiating Device. A device in which the sensing element is not ordinarily destroyed in the process of operation, whose restoration can be manual or automatic. **[72]**

Runner. A person other than the required number of operators on duty at central, supervising, or runner stations (or otherwise in contact with these stations) available for prompt dispatching, when necessary, to the protected premises. **[72]**

Runner Service. The service provided by a runner at the protected premises, including resetting and silencing of all equipment transmitting fire alarm or supervisory signals to an off-premises location. **[72]**

Satellite Trunk. A circuit or path connecting a satellite to its central or proprietary supervising station. **[72]**

Scanner. Equipment located at the telephone company wire center that monitors each local leg and relays status changes to the alarm center. Processors and associated equipment might also be included. **[72]**

Secondary Trunk Facility. That part of a transmission channel connecting two or more, but fewer than all, leg facilities to a primary trunk facility. **[72]**

Separate Sleeping Area. An area of the family living unit in which the bedrooms (or sleeping rooms) are located. Bedrooms (or sleeping rooms) separated by other use areas, such as kitchens or living rooms (but not bathrooms), are considered as separate sleeping areas. **[72]**

Shall. Indicates a mandatory requirement. **[72]**

Shapes of Ceilings. The shapes of ceilings can be classified as sloping or smooth. **[72]**

Should. Indicates a recommendation or that which is advised but not required. [72]

Signal. A status indication communicated by electrical or other means. [72]

Signal Transmission Sequence. A DACT that obtains dial tone, dials the number(s) of the DACR, obtains verification that the DACR is ready to receive signals, transmits the signals, and receives acknowledgment that the DACR has accepted that signal before disconnecting (going on-hook). [72]

Signaling Line Circuit. A circuit or path between any combination of circuit interfaces, control units, or transmitters over which multiple system input signals or output signals, or both, are carried. [72]

Signaling Line Circuit Interface. A system component that connects a signaling line circuit to any combination of initiating devices, initiating device circuits, notification appliances, notification appliance circuits, system control outputs, and other signaling line circuits. [72]

Single Dwelling Unit. A building consisting solely of one dwelling unit. *(See Dwelling Unit.)* [72]

Single Station Alarm. A detector comprising an assembly that incorporates a sensor, control components, and an alarm notification appliance in one unit operated from a power source either located in the unit or obtained at the point of installation. [72]

Single Station Alarm Device. An assembly that incorporates the detector, the control equipment, and the alarm-sounding device in one unit operated from a power supply either in the unit or obtained at the point of installation. [72]

Site-Specific Software. Software that defines the specific operation and configuration of a particular system. Typically, it defines the type and quantity of hardware modules, customized labels, and specific operating features of a system. [72]

Sloping Ceiling. A ceiling that has a slope of more than 1 in 8. **[72]**

Sloping Peaked-Type Ceiling. A ceiling in which the ceiling slopes in two directions from the highest point. Curved or domed ceilings can be considered peaked with the slope figured as the slope of the chord from highest to lowest point. **[72]**

Sloping Shed-Type Ceiling. A ceiling in which the high point is at one side with the slope extending toward the opposite side. **[72]**

Smoke Alarm. A single or multiple station alarm responsive to smoke. **[72]**

Smoke Detector. A device that detects visible or invisible particles of combustion. **[72]**

Smooth Ceiling. A ceiling surface uninterrupted by continuous projections, such as solid joists, beams, or ducts, extending more than 4 in. (100 mm) below the ceiling surface. **[72]**

Spacing. A horizontally measured dimension related to the allowable coverage of fire detectors. **[72]**

Spark. A moving ember. **[72]**

Spark/Ember Detector. A radiant energy-sensing fire detector that is designed to detect sparks or embers, or both. These devices are normally intended to operate in dark environments and in the infrared part of the spectrum. **[72]**

Spark/Ember Detector Sensitivity. The number of watts (or the fraction of a watt) of radiant power from a point source radiator, applied as a unit step signal at the wavelength of maximum detector sensitivity, necessary to produce an alarm signal from the detector within the specified response time. **[72]**

Spot-Type Detector. A device in which the detecting element is concentrated at a particular location. Typical examples are bimetallic detectors, fusible alloy detectors, certain pneumatic rate-of-rise detectors, certain smoke detectors, and thermoelectric detectors. [72]

Story. The portion of a building included between the upper surface of a floor and the upper surface of the floor or roof next above. [72]

Stratification. The phenomenon where the upward movement of smoke and gases ceases due to the loss of buoyancy. [72]

Structure, Open-Air Parking. A structure used for the parking or storage of motor vehicles that have (1) uniformly distributed openings in exterior walls on not less than two sides totaling not less than 40 percent of the building perimeter, (2) aggregate areas of such openings in exterior walls in each level not less than 20 percent of the total perimeter wall area of each level, and (3) interior wall lines and columns not less than 20 percent open with openings distributed to allow ventilation. [101]

Subscriber. The recipient of a contractual supervising station signal service(s). In case of multiple, noncontiguous properties having single ownership, the term refers to each protected premises or its local management. [72]

Subsidiary Station. A subsidiary station is a normally unattended location that is remote from the supervising station and is linked by a communications channel(s) to the supervising station. Interconnection of signals on one or more transmission channels from protected premises with a communications channel(s) to the supervising station is performed at this location. [72]

Supervising Station. A facility that receives signals and at which personnel are in attendance at all times to respond to these signals. [72]

Supervisory Service. The service required to monitor performance of guard tours and the operative condition of fixed suppression systems or other systems for the protection of life and property. [72]

FIRE ALARM–RELATED DEFINITIONS

Supervisory Signal. A signal indicating the need for action in connection with the supervision of guard tours, the fire suppression systems or equipment, or the maintenance features of related systems. [72]

Supervisory Signal Initiating Device. An initiating device such as a valve supervisory switch, water level indicator, or low air pressure switch on a dry-pipe sprinkler system in which the change of state signals an off-normal condition and its restoration to normal of a fire protection or life safety system; or a need for action in connection with guard tours, fire suppression systems or equipment, or maintenance features of related systems. [72]

Supplementary. As used in this code, supplementary refers to equipment or operations not required by this code and designated as such by the authority having jurisdiction. [72]

Switched Telephone Network. An assembly of communications facilities and central office equipment operated jointly by authorized service providers that provides the general public with the ability to establish transmission channels via discrete dialing. [72]

System Unit. The active subassemblies at the central station used for signal receiving, processing, display, or recording of status change signals; a failure of one of these subassemblies causes the loss of a number of alarm signals by that unit. [72]

Tactile Notification Appliance. A notification appliance that alerts by the sense of touch or vibration. [72]

Textual Audible Notification Appliance. A notification appliance that conveys a stream of audible information. An example of a textual audible notification appliance is a speaker that reproduces a voice message. [72]

Textual Visible Notification Appliance. A notification appliance that conveys a stream of visible information that displays an alphanumeric or pictorial message. Textual visible notification appliances provide temporary text, permanent text, or symbols. Textual visible notification appliances include, but are not limited to, annunciators, monitors, CRTs, displays, and printers. **[72]**

Tower. An enclosed independent structure or portion of a building with elevated levels for support of equipment or occupied for observation, control, operation, signaling, or similar limited use where (1) the elevated levels are provided to allow adequate observation or line-of-sight for personnel or equipment, and (2) the levels within the tower below the observation level and equipment room for that level are not occupied. **[101]**

Transmission Channel. A circuit or path connecting transmitters to supervising stations or subsidiary stations on which signals are carried. **[72]**

Transmitter. A system component that provides an interface between signaling line circuits, initiating device circuits, or control units and the transmission channel. **[72]**

Transponder. A multiplex alarm transmission system functional assembly located at the protected premises. **[72]**

Trouble Signal. A signal initiated by the fire alarm system or device indicative of a fault in a monitored circuit or component. **[72]**

Visible Notification Appliance. A notification appliance that alerts by the sense of sight. **[72]**

Voice Intelligibility. Audible voice information that is distinguishable and understandable. **[72]**

WATS (Wide Area Telephone Service). Telephone company service allowing reduced costs for certain telephone call arrangements. In-WATS or 800-number service calls can be placed from anywhere in the continental United States to the called party at no cost to the calling party. Out-WATS is a service whereby, for a flat-rate charge, dependent on the total duration of all such calls, a subscriber can make an unlimited number of calls within a prescribed area from a particular telephone terminal without the registration of individual call charges. **[72]**

Wavelength. The distance between the peaks of a sinusoidal wave. All radiant energy can be described as a wave having a wavelength. Wavelength serves as the unit of measure for distinguishing between different parts of the spectrum. Wavelengths are measured in microns (µM), nanometers (nM), or angstroms (Å). **[72]**

Wireless Control Panel. A component that transmits/receives and processes wireless signals. **[72]**

Wireless Protection System. A system or a part of a system that can transmit and receive signals without the aid of wire. It can consist of either a wireless control panel or a wireless repeater. **[72]**

Wireless Repeater. A component used to relay signals between wireless receivers or wireless control panels, or both. **[72]**

Zone. A defined area within the protected premises. A zone can define an area from which a signal can be received, an area to which a signal can be sent, or an area in which a form of control can be executed. **[72]**

EMERGENCY SYSTEMS DEFINITIONS

Editor's Note: The definitions in this section have been extracted from Article 700 of NFPA 70, National Electrical Code®, 2002 edition.

Emergency Systems. Emergency systems are those systems legally required and classed as emergency by municipal, state, federal, or other codes, or by any governmental agency having jurisdiction. These systems are intended to automatically supply illumination, power, or both, to designated areas and equipment in the event of failure of the normal supply or in the event of an accident to elements of a system intended to supply, distribute, and control power and illumination essential for safety to human life.

EMERGENCY SYSTEMS DEFINITIONS

Legally Required Standby Systems. Those systems required and so classed as legally required standby by municipal, state, federal, or other codes or by any governmental agency having jurisdiction. These systems are intended to automatically supply power to selected loads (other than those classed as emergency systems) in the event of failure of the normal source.

Optional Standby Systems. Those systems intended to protect public or private facilities or property where life safety does not depend on the performance of the system. Optional standby systems are intended to supply on-site generated power to selected loads either automatically or manually.

SECTION II

FUNDAMENTALS

This section provides a basic overview of the requirements for a fire alarm control unit, alarm initiation, notification, interfaces to other building systems, and miscellaneous components commonly used in the design of fire protection systems. The tables and figures in this section are essential to engineers, designers, contractors, and installers involved in designing and installing systems and verifying design requirements.

In addition to system types, power supply requirements, battery calculations, circuit class and style, monitoring for integrity, circuits extending beyond one building, alarm sequence, control functions, signals and survivability, the tables and figures provide information needed to determine the total system effectiveness.

Exhibit II.1 Fire Alarm Signaling Systems

Type	Description	Comments
1. Protected premises "local" fire alarm system	An alarm system operating in the protected premises, responsive to the operation of a manual fire alarm box, waterflow in a sprinkler system, or detection of a fire by a smoke or heat detection system.	The main purpose of this system to provide an evacuation or relocation alarm for the occupants of the building. Someone must always be present to transmit the alarm to fire authorities. See NFPA 72®, National Fire Alarm Code®.
2. Auxiliary fire alarm system	An alarm system utilizing a standard municipal coded fire alarm box to transmit a fire alarm from a protected premises to municipal fire headquarters. These alarms are received on the same municipal equipment and are carried over the same transmission lines (i.e. Public Fire Alarm Reporting System) as are used to connect fire alarm boxes located on streets. Operation is initiated by the local fire detection and alarm system installed at the protected premises.	Direct means of summoning help from municipal fire department. Some communities will accept this type of system and others will not. See NFPA 72, National Fire Alarm Code. Trouble signal may register in a separate attended location.
3. Remote supervising station fire alarm system	An alarm system connecting protected premises to a remote station, such as a fire station or a police station. Includes separate receiver for individual functions being monitored, such as fire alarm signal, or sprinkler waterflow alarm.	Requires connection into each protected premises. See NFPA 72, National Fire Alarm Code.

4.	Proprietary supervisory station fire alarm system	An alarm system that serves contiguous or noncontiguous properties under one ownership from a central supervising station. Similar to a central station system but owned by the protected property.	Requires 24 hr manning of proprietary supervising system station. See *NFPA 72, National Fire Alarm Code.*
5.	Central station fire alarm system	An alarm system connecting protected premises to a privately owned central station whose function is to monitor the connecting lines constantly and record any indication of fire, supervisory or other trouble signals from the protected premises. When a signal is received, the central station will take such action as is required, such as informing the municipal fire department of a fire or notifying the police department of intrusion.	Flexible system. Can handle many types of alarms, including trouble within system at protected premises. See *NFPA 72, National Fire Alarm Code.*
6.	Emergency voice/ alarm communications system	Provides for the inclusion of emergency voice/ alarm communications.	Provides dedicated facilities for the transmission of information to occupants of the building (including fire department personnel). See *NFPA 72, National Fire Alarm Code.*

Source: Richard W. Bukowski and Robert J. O'Laughlin, *Fire Alarm Signaling Systems,* 2nd ed., NFPA, Quincy, MA; Society of Fire Protection Engineers, Boston, 1994, Table 2-1.

Exhibit II.2 Fire Extinguishing Systems

Type	Description	Comments
1. Wet-pipe automatic sprinkler system	A permanently piped water system under pressure, using heat-actuated sprinklers. When a fire occurs, the sprinklers exposed to the high heat open and discharge water individually to control or extinguish the fire.	Automatically detects and controls fire. Protects structure. May cause water damage to unprotected books, manuscripts, records, paintings, specimens, or other valuable objects. Not to be used in spaces subject to freezing. On-off types may limit water damage. See NFPA 13, *Standard for the Installation of Sprinkler Systems*, and NFPA 22, *Standard for Water Tanks for Private Fire Protection.*
2. Pre-action automatic sprinkler system	A system employing automatic sprinklers attached to a piping system containing air that may or may not be under pressure, with a supplemental fire detection system installed in the same area as the sprinklers. Actuation of the fire detection system by a fire opens a valve that permits water to flow into the sprinkler system piping and to be discharged from any sprinklers that are opened by the heat from the fire.	Automatically detects and controls fire. May be installed in areas subject to freezing. Minimizes the accidental discharge of water due to mechanical damage to sprinkler heads or piping. See NFPA 13, *Standard for the Installation of Sprinkler Systems*, and NFPA 22, *Standard for Water Tanks for Private Fire Protection.*

3. On-off automatic sprinkler system

A system similar to the pre-action system, except that the fire detector operation acts as an electrical interlock, causing the control valve to open at a predetermined temperature and close when normal temperature is restored. Should the fire rekindle after its initial control, the valve will reopen and water will again flow from the opened heads. The valve will continue to open and close in accordance with the temperature sensed by the fire detectors. Another type of on-off system is a standard wet-pipe system with on-off sprinkler heads. Here, each individual head has incorporated in it a temperature-sensitive device that causes the head to open at a predetermined temperature and close automatically when the temperature at the head is restored to normal.

4. Dry-pipe automatic sprinkler system

A system that has heat-operated sprinklers attached to a piping system containing air under pressure. When a sprinkler operates, the air pressure is reduced, a "dry-pipe" valve is opened by water pressure, and water flows to any opened sprinklers.

In addition to the favorable feature of the automatic wet-pipe system, these systems have the ability to automatically stop the flow of water when no longer needed, thus eliminating unnecessary water damage. See NFPA 13, *Standard for the Installation of Sprinkler Systems,* and NFPA 22, *Standard for Water Tanks for Private Fire Protection.*

See No. 1. Can protect areas subject to freezing. Water supply must be in a heated area. See NFPA 13, *Standard for the Installation of Sprinkler Systems,* and NFPA 22, *Standard for Water Tanks for Private Fire Protection.*

Continued

Exhibit II.2 Fire Extinguishing Systems *Continued*

Type	Description	Comments
5. Standpipe and hose system	A piping system in a building to which hoses are connected for emergency use by building occupants or by the fire department.	A desirable complement to an automatic sprinkler system. Staff requires training to use hoses effectively. See NFPA 14, *Standard for the Installation of Standpipe, Private Hydrant, and Hose Systems.*
6. Halon and halon alternatives automatic system	A permanently piped system using a limited stored supply of a halon (or halon alternative) gas under pressure, and discharge nozzles which can totally flood an enclosed area. Released automatically by a suitable detection system. Extinguishes fires by inhibiting the chemical reaction of fuel and oxygen.	No agent damage to unprotected books, manuscripts, records, paintings, or other irreplaceable valuable objects. No agent residue. Halon 1301 can be used with safeguards in normally occupied areas. Halon 1211 total flooding systems are prohibited in normally occupied areas. Halons may not extinguish deep-seated fires in ordinary solid combustibles, such as paper, fabrics, etc.; but are effective on surface fires in these materials. These systems require special precautions to avoid damage effects caused by their extremely rapid release. The high-velocity discharge from nozzles may be sufficient to dislodge substantial objects directly in the path. See NFPA 12A, *Standard on Halon 1301 Fire Extinguishing Systems* and NFPA 2001, *Standard on Clean Agent Fire Extinguishing Systems.*

7. Carbon dioxide automatic system

Same as No. 6, except uses carbon dioxide gas. Extinguishes fires by reducing oxygen content of air below combustion support point.

Same as No. 6. Appropriate for service and utility areas. Personnel must evacuate before agent discharge to avoid suffocation. May not extinguish deep-seated fires in ordinary solid combustibles, such as paper, fabrics, etc.; but effective on surface fires in these materials. See NFPA 12, *Standard on Carbon Dioxide Extinguishing System.*

8. Dry chemical automatic system

Same as No. 6, except uses a dry chemical powder. Usually released by mechanical thermal linkage. Effective for surface protection.

Should not be used in personnel-occupied areas. Leaves powdery deposit on all exposed surfaces. Requires cleanup. Excellent for service facilities having kitchen range hoods and ducts. May not extinguish deep-seated fires in ordinary solid combustibles, such as paper, fabrics, etc.; but effective on surface fires in these materials. See NFPA 17, *Standard for Dry Chemical Extinguishing Systems.*

Continued

Exhibit II.2 Fire Extinguishing Systems *Continued*

Type	Description	Comments
9. High-expansion foam system	A fixed extinguishing system that generates a foam agent for total flooding of confined spaces, and for volumetric displacement of vapor, heat, and smoke. Acts on the fire by: a. Preventing free movement of air. b. Reducing the oxygen concentration at the fire. c. Cooling.	Should not be used in occupied areas. The discharge of large amounts of high-expansion foam may inundate personnel, blocking vision, making hearing difficult, and creating some discomfort in breathing. Leaves residue and requires cleanup. High-expansion foam when used in conjunction with water sprinklers will provide more positive control and extinguishment than either extinguishment system used independently, when properly designed. See NFPA 11A, *Standard for Medium- and High-Expansion Foam Systems.*

*Source: Modified from Richard W. Bukowski and Robert J. O'Laughlin, *Fire Alarm Signaling Systems*, 2nd ed., NFPA, Quincy, MA; Society of Fire Protection Engineers, Boston, 1994, Table 3-3.*

Exhibit II.3 Primary Power Supply Requirements

Component	Requirement
Primary supply	High degree of reliability
	Adequate capacity
	Must be either:
	1. Light and power service
	2. Engine-driven generator
Branch circuit	Dedicated to the fire alarm system
Branch circuit wiring	Mechanically protected
Branch circuit connections	Mechanically protected
Disconnecting means	Red marking
	Accessible only to authorized persons
	Identified as "FIRE ALARM CIRCUIT CONTROL"
Overcurrent protection	Locate in a locked or sealed cabinet located immediately adjacent to the point of connection to the light and power conductors

Exhibit II.4 Secondary Power Supply Requirements

System Type	Maximum Quiescent Load	Total Alarm Load
Protected premises system	24 hr	5 min
Central station system	24 hr	5 min
Auxiliary system	60 hr	5 min
Remote station system	60 hr	5 min
Proprietary system	24 hr	5 min
Emergency voice/alarm communication system	24 hr	15 min[a]

[a]All connected notification appliances

Source: Modified from Richard W. Bukowski and Robert J. O'Laughlin, *Fire Alarm Signaling Systems,* 2nd ed., NFPA, Quincy, MA; Society of Fire Protection Engineers, Boston, 1994, Table 2-2.

POWER SUPPLIES

Exhibit II.5 Power Supply Sources

Primary Supply	Secondary Supply
Light and power service	Batteries
Engine-driven generator (where a person specifically trained in its operation is on duty at all times)	Automatic starting engine-driven generator with 4 hr of battery capacity
	Multiple engine-driven generators

Editor's Note: Where system is powered from a dedicated branch circuit of an Emergency System, Legally Required Standby System, or an Optional Standby System that meets requirements of an Emergency or Legally Required Standby System, no secondary power supply is required.

Exhibit II.6 Nominal Battery Voltages and Specific Gravities

Battery Type	Nominal Cell Voltage at 25°C (77°F) (when connected to charger)	Specific Gravity Range
Vented lead acid, gelled or starved electrolyte	2.30 ± 0.02	1.205–1.220 (regular performance) 1.240–1.260 (high performance)
Nickel cadmium	1.42	N/A
Sealed lead acid	2.30 ± 0.02	N/A

Editor's Note: Batteries to be located so that fire alarm equipment is not adversely affected by battery gases and per NFPA 70, National Electrical Code®, Article 480. If batteries are not located in or adjacent to the fire alarm control panel, the batteries and their charger location must be permanently identified at the fire alarm control unit.

Exhibit II.7 Battery Calculation Sheet

Item	Description	Standby Current per Unit (Amps)		Qty		Total System Standby Current per Item (Amps)	Alarm Current per Unit (Amps)		Qty		Total System Alarm Current per Item (Amps)
			×		=			×		=	
			×		=			×		=	
			×		=			×		=	
			×		=			×		=	
			×		=			×		=	
			×		=			×		=	
			×		=			×		=	
					Total System Standby Current (Amps)					Total System Alarm Current (Amps)	

Continued

Exhibit II.7 Battery Calculation Sheet *Continued*

Maximum Quiescent Load (Standby): _____ hr   Total Alarm Load: _____ min × 1/60 = _____ hr

Required Standby Time (hr)		Total System Standby Current (Amps)		Required Standby Capacity (Amp-hr)		Required Alarm Time (hr)		Total System Alarm Current (Amps)		Required Alarm Capacity (Amp-hr)
	×		=				×		=	

Required Standby Capacity (Amp-hr)		Required Alarm Capacity (Amp-hr)		Total Required Capacity (Amp-hr)		Optional Factor of Safety		Required Battery Capacity (Amp-hr)
	+		=			×	=	

Exhibit II.8 Typical Performance Characteristics of Lead Acid Batteries

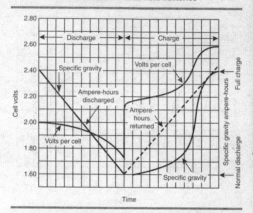

Source: Donald G. Fink and H. Wayne Beaty, eds., *Standard Handbook for Electrical Engineers*, 12th ed., McGraw-Hill, New York, 1987, Fig. 11-56.

Editor's Note: During discharge, the specific gravity decreases from a fully charged to a fully discharged condition.

Exhibit II.9 Discharge Curves of Lead Acid Batteries at Different Hour Rates

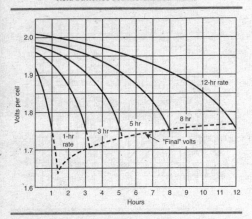

Source: Donald G. Fink and H. Wayne Beaty, eds., *Standard Handbook for Electrical Engineers*, 12th ed., McGraw-Hill, New York, 1987, Fig. 11-57.

Editor's Note: Nominal cell voltage on open circuit with battery fully charged is 2.00 to 2.12 V.

Exhibit II.10 Performance of Lead Acid Batteries at Various Temperatures

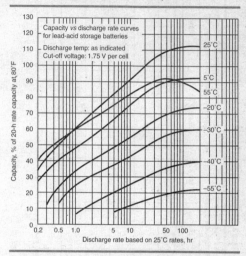

Source: Donald G. Fink and H. Wayne Beaty, eds., *Standard Handbook for Electrical Engineers*, 12th ed., McGraw-Hill, New York, 1987, Fig. 11-59.

Editor's Note: The battery will operate under a wide temperature range, but higher service capacity is obtained at lower discharge rates and higher temperatures.

Exhibit II.11 Class 2 and Class 3 Alternating Current Power Source Limitations

Power Source		Inherently Limited Power Source (Overcurrent Protection Not Required)				Not Inherently Limited Power Source (Overcurrent Protection Required)			
		Class 2			Class 3	Class 2			Class 3
		0 through 20ᵃ	Over 20 and through 30ᵃ	Over 30 and through 150	Over 30 and through 100	0 through 20ᵃ	Over 20 and through 30ᵃ	Over 30 and through 100	Over 100 and through 150
Source voltage V_{max} (volts) (see Note 1)		0 through 20ᵃ	Over 20 and through 30ᵃ	Over 30 and through 150	Over 30 and through 100	0 through 20ᵃ	Over 20 and through 30ᵃ	Over 30 and through 100	Over 100 and through 150
Power limitations VA_{max} (volt-amperes) (see Note 1)		—	—	—	—	250 (see Note 3)	250	250	N.A.
Current limitations I_{max} (amperes) (see Note 1)		8.0	8.0	0.005	150/V_{max}	1000/V_{max}	1000/V_{max}	1000/V_{max}	1.0
Maximum overcurrent protection (amperes)		—	—	—	—	5.0	100/V_{max}	100/V_{max}	1.0
Power source maximum nameplate rating	VA (volt-amperes)	5.0 × V_{max}	100	0.005 × V_{max}	100	5.0 × V_{max}	100	100	100
	Current (amperes)	5.0	100/V_{max}	0.005	100/V_{max}	5.0	100/V_{max}	100/V_{max}	100/V_{max}

a Voltage ranges shown are for sinusoidal ac in indoor locations or where wet contact is not likely to occur. For nonsinusoidal or wet contact conditions, see Note 2.

Notes:

1. V_{max}, I_{max}, and VA_{max} are determined with the current limiting impedance in the circuit (not bypassed) as follows.

   V_{max}: Maximum output current under any noncapacitive load, including short circuit, and with overcurrent protection bypassed if used. Where a transformer limits the output current. I_{max} limits apply voltage regardless of load with rated input applied.

   I_{max}: Maximum output after 1 min of operation. Where a current-limiting impedance, listed for the purpose, or as part of a listed product, is used in combination with a nonpower-limited transformer or a stored energy source, e.g., storage battery, to limit the output current, I_{max} limits apply after 5 s.

   VA_{max}: Maximum volt-ampere output after 1 min of operation regardless of load and overcurrent protection bypassed if used.

2. For nonsinusoidal ac, V_{max} shall not be greater than 42.4 V peak. Where wet contact (immersion not included) is likely to occur, Class 3 wiring methods shall be used or V_{max} shall not be greater than 15 V for sinusoidal ac and 21.2 V peak for nonsinusoidal ac.

3. If the power source is a transformer, VA_{max} is 350 or less where V_{max} is 15 or less.

4. For dc interrupted at a rate of 10 to 200 Hz, V_{max} shall not be greater than 24.8 V peak. Where wet contact (immersion not included) is likely to occur, Class 3 wiring methods shall be used or V_{max} shall not be greater than 30 V for continuous dc; 12.4 V peak for dc that is interrupted at a rate of 10 to 200 Hz.

Source: NFPA 70, *National Electrical Code®*, NFPA, Quincy, MA, 2002 edition, Table 11(a).

Exhibit II.12 Class 2 and Class 3 Direct Current Power Source Limitations

Power Source	Inherently Limited Power Source (Overcurrent Protection Not Required)					Not Inherently Limited Power Source (Overcurrent Protection Required)			
	Class 2			Class 3		Class 2		Class 3	
Source voltage V_{max} (volts) (see Note 1)	0 through 20[a]	Over 20 and through 30[a]	Over 30 and through 60[a]	Over 60 and through 100	Over 60 and through 150	0 through 20[a]	Over 20 and through 60[a]	Over 60 and through 100	Over 100 and through 150
Power limitations VA_{max} (volt-amperes) (see Note 1)	—	—	—	—	—	250 (see Note 3)	250	250	N.A.
Current limitations I_{max} (amperes) (see Note 1)	8.0	8.0	150/V_{max}	150/V_{max}	0.005	1000/V_{max}	1000/V_{max}	1000/V_{max}	1.0
Maximum overcurrent protection (amperes)	—	—	—	—	—	5.0	100/V_{max}	1.0	1.0
Power source maximum nameplate rating — VA (volt-amperes)	5.0 × V_{max}	100	100	100	0.005 × V_{max}	5.0 × V_{max}	100	100	100
Power source maximum nameplate rating — Current (amperes)	5.0	100/V_{max}	100/V_{max}	100/V_{max}	0.005	5.0	100/V_{max}	100/V_{max}	100/V_{max}

[a]Voltage ranges shown are for continuous dc in indoor locations or where wet contact is not likely to occur. For interrupted dc or wet contact conditions, see Note 4.

Notes:

1. V_{max}, I_{max}, and VA_{max} are determined with the current limiting impedance in the circuit (not bypassed) as follows.

   V_{max}: Maximum output current under any noncapacitive load; including short circuit, and with overcurrent protection bypassed if used. Where a transformer limits the output current, I_{max} limits apply voltage regardless of load with rated input applied.

   I_{max}: Maximum output after 1 min of operation. Where a current-limiting impedance, listed for the purpose, or as part of a listed product, is used in combination with a nonpower-limited transformer or a stored energy source, e.g., storage battery, to limit the output current, I_{max} limits apply after 5 s.

   VA_{max}: Maximum volt-ampere output after 1 min of operation regardless of load and overcurrent protection bypassed if used.

2. For nonsinusoidal ac, V_{max} shall not be greater than 42.4 V peak. Where wet contact (immersion not included) is likely to occur, Class 3 wiring methods shall be used or V_{max} shall not be greater than 15 V for sinusoidal ac and 21.2 V peak for nonsinusoidal ac.

3. If the power source is a transformer, VA_{max} is 350 or less when V_{max} is 15 or less.

4. For dc interrupted at a rate of 10 to 200 Hz, V_{max} shall not be greater than 24.8 V peak. Where wet contact (immersion not included) is likely to occur, Class 3 wiring methods shall be used or V_{max} shall not be greater than 30 V for continuous dc; 12.4 V peak for dc that is interrupted at a rate of 10 to 200 Hz.

Source: NFPA 70, National Electrical Code®, NFPA, Quincy, MA, 2002 edition, Table 11(b).

Exhibit II.13 PLFA Alternating Current Power Source Limitations

Power Source	Inherently Limited Power Source (Overcurrent Protection Not Required)			Not Inherently Limited Power Source (Overcurrent Protection Required)		
Circuit voltage V_{max} (volts) (see Note 1)	0 through 20	Over 20 and through 30	Over 30 and through 100	0 through 20	Over 20 and through 100	Over 100 and through 150
Power limitations VA_{max} (volt-amperes) (see Note 1)	—	—	—	250 (see Note 2)	250	N.A.
Current limitations I_{max} (amperes) (see Note 1)	8.0	8.0	150/V_{max}	1000/V_{max}	1000/V_{max}	1.0
Maximum overcurrent protection (amperes)	—	—	—	5.0	100/V_{max}	1.0
Power source maximum nameplate ratings — VA (volt-amperes)	5.0 × V_{max}	100	100	5.0 × V_{max}	100	100
Power source maximum nameplate ratings — Current (amperes)	5.0	100/V_{max}	100/V_{max}	5.0	100/V_{max}	100/V_{max}

Notes:

1. V_{max}, I_{max}, and VA_{max} are determined as follows.

V_{max}: Maximum output voltage regardless of load with rated input applied.

I_{max}: Maximum output current under any noncapacitive load, including short circuit, and with overcurrent protection bypassed if used. Where a transformer limits the output current, I_{max} limits apply after 1 min of operation. Where a current-limiting impedance, listed for the purpose, is used in combination with a nonpower-limited transformer or a stored energy source, e.g., storage battery, to limit the output current, I_{max} limits apply after 5 s.

VA_{max}: Maximum volt-ampere output after 1 min of operation regardless of load and overcurrent protection bypassed if used. Current limiting impedance shall not be bypassed when determining I_{max} and VA_{max}.

2. If the power source is a transformer, VA_{max} is 350 or less when V_{max} is 15 or less.

Source: NFPA 70, National Electrical Code®, NFPA, Quincy, MA, 2002 edition, Table 12(a).

Exhibit II.14 PLFA Direct Current Power Source Limitations

Power Source		Inherently Limited Power Source (Overcurrent Protection Not Required)				Not Inherently Limited Power Source (Overcurrent Protection Required)		
		0 through 20	Over 20 and through 30	Over 30 and through 100	Over 100 and through 250	0 through 20	Over 20 and through 100	Over 100 and through 150
Circuit voltage V_{max} (volts) (see Note 1)		0 through 20	Over 20 and through 30	Over 30 and through 100	Over 100 and through 250	0 through 20	Over 20 and through 100	Over 100 and through 150
Power limitations VA_{max} (volt-amperes) (see Note 1)		—	—	—	—	250 (see Note 2)	250	N.A.
Current limitations I_{max} (amperes) (see Note 1)		8.0	8.0	$150/V_{max}$	0.030	$1000/V_{max}$	$1000/V_{max}$	1.0
Maximum overcurrent protection (amperes)		—	—	—	—	5.0	$100/V_{max}$	1.0
Power source maximum nameplate ratings	VA (volt-amperes)	$5.0 \times V_{max}$	100	100	$0.030 \times V_{max}$	$5.0 \times V_{max}$	100	100
	Current (amperes)	5.0	$100/V_{max}$	$100/V_{max}$	0.030	5.0	$100/V_{max}$	$100/V_{max}$

Notes:

1. V_{max}, I_{max}, and VA_{max} are determined as follows.

   V_{max}: Maximum output voltage regardless of load with rated input applied.

   I_{max}: Maximum output current under any noncapacitive load, including short circuit, and with overcurrent protection bypassed if used. Where a transformer limits the output current, I_{max} limits apply after 1 min of operation. Where a current-limiting impedance, listed for the purpose, is used in combination with a nonpower-limited transformer or a stored energy source, e.g., storage battery, to limit the output current, I_{max} limits apply after 5 s.

   VA_{max}: Maximum volt-ampere output after 1 min of operation regardless of load and overcurrent protection bypassed if used. Current limiting impedance shall not be bypassed when determining I_{max} and VA_{max}.

2. If the power source is a transformer, VA_{max} is 350 or less when V_{max} is 15 or less.

Source: NFPA 70, *National Electrical Code®*, NFPA, Quincy, MA, 2002 edition, Table 12(b).

Exhibit II.15 Performance of Initiating Device Circuits (IDC)

| | Class B | | | Class B | | |
| | Style A | | | Style B | | |
Abnormal Condition	Alarm	Trouble	Alarm Receipt Capability during Abnormal Condition	Alarm	Trouble	Alarm Receipt Capability during Abnormal Condition
	1	2	3	4	5	6
Single open	—	X	—	—	X	—
Single ground	—	X	—	—	X	R
Wire-to-wire short	X	—	—	X	—	—
Loss of carrier (if used)/ channel interface	—	—	—	—	—	—

Abnormal Condition	Class B Style C			Class A Style D		
	Alarm	Trouble	Alarm Receipt Capability during Abnormal Condition	Alarm	Trouble	Alarm Receipt Capability during Abnormal Condition
	7	8	9	10	11	12
Single open	—	X	—	—	X	X
Single ground	—	X	R	—	X	R
Wire-to-wire short	—	X	—	X	—	—
Loss of carrier (if used)/channel interface	—	X	—	—	—	—

Continued

Exhibit II.15 Performance of Initiating Device Circuits (IDC) *Continued*

	Class A		
	Style Eα		
			Alarm Receipt Capability during Abnormal Condition
	Alarm	Trouble	
Abnormal Condition	13	14	15
Single open	—	X	X
Single ground	—	X	R
Wire-to-wire short	—	X	—
Loss of carrier (if used)/channel interface	—	X	—

R = Required capacity

X = Indication required at protected premises and as required by Chapter 5 of *NFPA 72*.

α = Style exceeds minimum requirements of Class A

Source: NFPA 72®, National Fire Alarm Code®, NFPA, Quincy, MA, 1999 edition, Table 3-5.

Exhibit II.16 Performance of Signaling Line Circuits (SLC)

	Class B			Class B		
	Style 0.5			Style 1		
Abnormal Condition	Alarm	Trouble	Alarm Receipt Capability during Abnormal Conditions	Alarm	Trouble	Alarm Receipt Capability during Abnormal Conditions
	1	2	3	4	5	6
Single open	—	X	—	—	X	—
Single ground	—	X	—	—	X	R
Wire-to-wire short	—	—	—	—	—	—
Wire-to-wire short and open	—	—	—	—	—	—
Wire-to-wire short and ground	—	—	—	—	—	—
Open and ground	—	—	—	—	—	—
Loss of carrier (if used)/channel interface	—	—	—	—	—	—

Continued

Exhibit II.16 Performance of Signaling Line Circuits (SLC) *Continued*

	Class A			Class B		
	Style 2α			Style 3		
Abnormal Condition	Alarm	Trouble	Alarm Receipt Capability during Abnormal Conditions	Alarm	Trouble	Alarm Receipt Capability during Abnormal Conditions
	7	8	9	10	11	12
Single open	—	X	R	—	X	R
Single ground	—	X	R	—	X	R
Wire-to-wire short	—	—	M	—	X	—
Wire-to-wire short and open	—	—	M	—	X	—
Wire-to-wire short and ground	—	X	M	—	X	—
Open and ground	—	X	R	—	X	—
Loss of carrier (if used)/channel interface	—	—	—	—	—	—

	Class B			Class B		
	Style 3.5			Style 4		
Abnormal Condition	Alarm	Trouble	Alarm Receipt Capability during Abnormal Conditions	Alarm	Trouble	Alarm Receipt Capability during Abnormal Conditions
	13	14	15	16	17	18
Single open	—	X	—	—	X	—
Single ground	—	X	—	—	X	R
Wire-to-wire short	—	X	—	—	X	—
Wire-to-wire short and open	—	X	—	—	X	—
Wire-to-wire short and ground	—	X	—	—	X	—
Open and ground	—	X	—	—	X	—
Loss of carrier (if used)/channel interface	—	X	—	—	X	—

Continued

Exhibit II.16 Performance of Signaling Line Circuits (SLC) *Continued*

Abnormal Condition	Class B Style 4.5			Class A Style 5α		
	Alarm	Trouble	Alarm Receipt Capability during Abnormal Conditions	Alarm	Trouble	Alarm Receipt Capability during Abnormal Conditions
	19	20	21	22	23	24
Single open	—	X	R	—	X	R
Single ground	—	X	R	—	X	R
Wire-to-wire short	—	X	—	—	X	—
Wire-to-wire short and open	—	X	—	—	X	—
Wire-to-wire short and ground	—	X	—	—	X	—
Open and ground	—	X	—	—	X	—
Loss of carrier (if used)/channel interface	—	X	—	—	X	—

Abnormal Condition	Class A Style 6α			Class A Style 7α		
	Alarm	Trouble	Alarm Receipt Capability during Abnormal Conditions	Alarm	Trouble	Alarm Receipt Capability during Abnormal Conditions
	25	26	27	28	29	30
Single open	—	X	R	—	X	R
Single ground	—	X	R	—	X	R
Wire-to-wire short	—	X	—	—	X	R
Wire-to-wire short and open	—	X	—	—	X	—
Wire-to-wire short and ground	—	X	—	—	X	R
Open and ground	—	X	X	—	X	—
Loss of carrier (if used)/channel interface	—	X	X	—	X	—

M = May be capable of alarm with wire-to-wire short

R = Required capability

X = Indication required at protected premises and as required by Chapter 5 of *NFPA 72*.

α = Style exceeds minimum requirements for Class A

Source: NFPA 72®, National Fire Alarm Code®, NFPA, Quincy, MA, 1999 edition, Table 3-6.

Exhibit II.17 Performance of Notification Appliance Circuits (NAC)

Abnormal Condition	Class B Style W		Class B Style X	
	Trouble Indication at Protected Premises	Alarm Capability during Abnormal Conditions	Trouble Indication at Protected Premises	Alarm Capability during Abnormal Conditions
	1	2	3	4
Single open	X	—	X	X
Single ground	X	—	X	—
Wire-to-wire short	X	—	X	—

Abnormal Condition	Class B Style Y		Class A Style Z	
	Trouble Indication at Protected Premises	Alarm Capability during Abnormal Conditions	Trouble Indication at Protected Premises	Alarm Capability during Abnormal Conditions
	5	6	7	8
Single open	X	—	X	X
Single ground	X	X	X	X
Wire-to-wire short	X	—	X	—

X = Indication required at protected premises

Source: NFPA 72®, National Fire Alarm Code®, NFPA, Quincy, MA, 1999 edition, Table 3-7.

Exhibit II.18 Separation of Class A Circuits

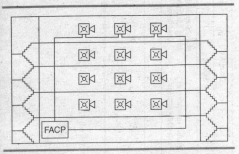

Editor's Note: Class A outgoing and return conductors exiting from and returning to the control unit are routed separately. The minimum separation is 1 ft when the cable is installed vertically and 4 ft when run horizontally.

**Exhibit II.19 Separation of Class A Circuits—
Installation Exception (a): Distance Not to Exceed 10 ft**

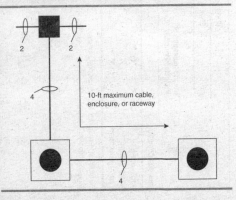

10-ft maximum cable, enclosure, or raceway

Exhibit II.20a Separation of Class A Circuits—Installation Exception (b): 2-hr Rated Enclosure

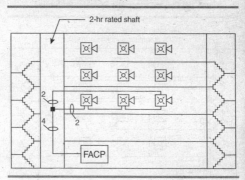

Exhibit II.20b Separation of Class A Circuits—Installation Exception (b): 2-hr Rated Cable Assembly

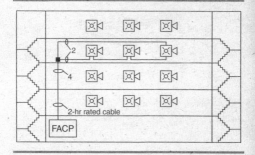

**Exhibit II.21 Separation of Class A Circuits—
Installation Exception: 2-hr Rated Stairwell
and 100 Percent Automatic Sprinkler Protection**

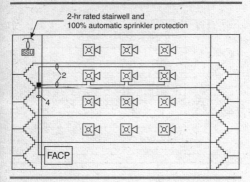

2-hr rated stairwell and
100% automatic sprinkler protection

Editor's Note: Some local codes may prohibit use of this exception.

Exhibit II.22 Separation of Class A Circuits—
Installation Exception (c): Unlimited Drops in Conduit
or Raceway to a Single Device or Appliance

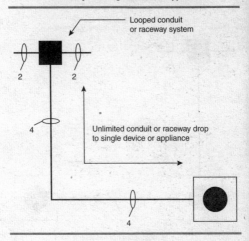

Looped conduit
or raceway system

Unlimited conduit or raceway drop
to single device or appliance

**Exhibit II.23 Separation of Class A Circuits—
Installation Exception (d): Unlimited Drops in Conduit or
Raceway to a Single Room Not Exceeding 1000 sq ft**

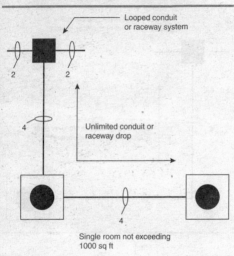

Looped conduit
or raceway system

2   2

4

Unlimited conduit or
raceway drop

4

Single room not exceeding
1000 sq ft

Exhibit II.24 Loading Capacities of Initiating Device Circuits

Type of Device	Maximum Quantity per IDC
Heat detector	Manufacturer specific
Smoke detector	Manufacturer specific
Radiant energy-sensing fire detector	Manufacturer specific
Other fire detectors	Manufacturer specific
Other automatic extinguishing devices	Manufacturer specific
Manual fire alarm box	Manufacturer specific
Waterflow device	5
Supervisory device	20

Editor's Note: Initiation of the waterflow alarm signal must occur within 90 s of waterflow at the alarm-initiating device when flow occurs that is equal to or greater than that from a single sprinkler of the smallest orifice size installed in the system. Piping between the sprinkler system and the pressure-activated alarm initiating device must be galvanized or of nonferrous metal or other approval corrosion-resistant material of not less than 3/8 in. nominal pipe size. If a valve is installed in the connection between an alarm-initiating device intended to signal activation of a fire suppression system and the fire suppression system, the valve must be supervised.

Exhibit II.25 Supervisory Conditions

Component	Supervisory Conditions	Trouble Conditions
Control valves	Position [1-1/2 in. (38.1 mm) or larger]	
Pressure switch	Dry-pipe system air Pressure tank air Preaction system air Steam for flooding systems Public water supplies	
Water tanks	Water level and temperature	
Buildings	Temperature	
Electric fire pumps	Pump running (alarm or supervisory) Power failure Phase reversal	
Engine-driven fire pumps	Engine running (alarm or supervisory) Failure to start Controller off "automatic"	Low oil High temp Overspeed
Steam turbine fire pumps	Pump running (alarm or supervisory) Steam pressure Steam control valves	

Editor's Note: Supervisory signals must be distinctive in sound from other signals. This sound cannot be used for any other purpose. They must distinctively indicate the particular function (e.g., valve position, temperature, or pressure).

Exhibit II.26 Fire Alarm Circuits Extending beyond One Building and Run Outdoors

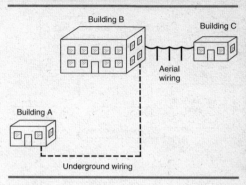

Editor's Note: See Exhibit II.27.

Exhibit II.27 Fire Alarm Circuits Extending beyond One Building Must Comply with *NEC* 760.7

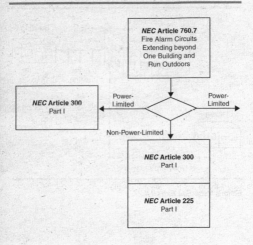

continued

Exhibit II.27 Fire Alarm Circuits Extending beyond One Building Must Comply with *NEC* 760.7 *Continued*

800.10 Overhead Communications Wires and Cables.
800.11 Underground Circuits Entering Buildings.
800.12 Circuits Requiring Primary Protectors
(As provided in 800.30).
800.13 Lightning Conductors.

NEC Article 800
Communications
Circuits

Part II. Conductors
Outside and
Entering Buildings.

Part III.
Protection.

Part IV. Grounding
Methods.

800.30 Protective Devices must be installed:
• Listed (See Article 285 TVSSs).
• On each circuit.
• One at each end of the circuit.

Exception:
(1) Circuits in large metropolitan areas where buildings are close together and sufficiently high to intercept lighting.
(2) Interbuilding cable runs of 140 ft or less, directly buried or in underground conduit, where a continuous metallic cable shield or a continuous metallic conduit containing the cable is bonded to each building grounding electrode system.
(3) Areas having an average of five or fewer thunderstorm days per year (See Section IX - 25) and earth resistivity of less than 100 ohm-meters.

800.31 Primary Protector Requirements.
800.32 Secondary Protector Requirements.
800.33 Cable Grounding.
Metallic sheath of communications cables to be grounded as close as practicable to the point of entrance.

800.40 Cable and Primary Protector Grounding.

(A) Grounding Conductor.
(1) Insulation. Listed for the purpose.
(2) Material. Copper or corrosion-resistant conductive material.
(3) Size. No. 14 minimum.
(4) Length. To be as short as practiceable.
(5) Run in Straight Line. Run to the grounding electrode in as straight a line as practicable.
(6) Physical Damage. If run in a metal raceway, both ends of the raceway must be bonded.

(B) Electrode.

800.41 Primary Protector Grounding and Bonding at Mobile Homes.

Exhibit II.28 Alarm Verification Flow Chart

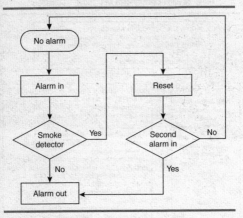

Exhibit II.29 Positive Alarm Sequence Flow Chart

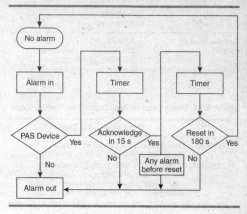

CONTROL FUNCTIONS

Exhibit II.30 Fire Safety Functions—Methods of Connections

Dry Contacts	Listed Digital Data Interfaces	Other
Contacts rated for the load	BACNet RS-232 RS-485	Listed methods permitted

Exhibit II.31 Smoke Detector Requirements for HVAC Systems

System CFM	Where Detectors Are Required
>2,000 CFM	Downstream of the fan and air filters and ahead of any branch connections in air supply systems.
>15,000 CFM	At each location prior to the connection to a common return and prior to any recirculation or fresh air inlet connection in air return systems serving more than one smoke compartment.

Exhibit II.32 Elevator Recall and Shutdown Requirements

Location	Type of Detector Required
Elevator lobbies	Smoke or other fire detector, depending on the ambient environment.
Elevator hoistways	Smoke and heat detection only if sprinklered. No detection if not sprinklered.
Elevator machine rooms	Smoke detection, if not sprinklered. If sprinklered, both smoke and heat detection.

Exhibit II.33 Elevator Recall System

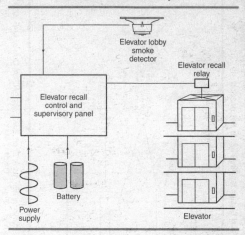

Source: Merton W. Bunker, Jr., and Wayne D. Moore, eds., *National Fire Alarm Code Handbook®*, NFPA, Quincy, MA, 1999, Exhibit 3.24. (Courtesy of FIREPRO Incorporated, Andover, MA).

Exhibit II.34 Connection to Third Circuit for Fire Fighter Notification

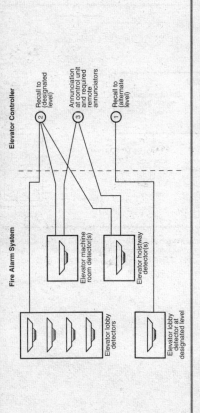

Source: Merton W. Bunker, Jr., and Wayne D. Moore, eds. *National Fire Alarm Code Handbook®*, NFPA, Quincy, MA, 1999, Exhibit 3.26 (Courtesy of Bruce Fraser, Simplex Time Recorder Company, Gardner, MA).

CONTROL FUNCTIONS

Exhibit II.35 Typical Method of Providing Elevator Power Shunt Trip Supervisory Signal

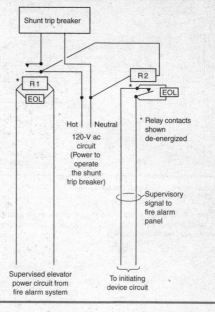

Source: NFPA 72®, National Fire Alarm Code®, NFPA, Quincy, MA, 1999 edition, Fig. A-3-9.4.4.

Editor's Note: Compared to the sprinkler, heat detectors used to shut down elevator power prior to sprinkler operation must have a lower temperature rating and a higher sensitivity. The heat detector, which is placed within 2 ft of each sprinkler, must shut down elevator power. The control circuit must be monitored for presence of operating voltage. The loss of voltage to the control circuit causes a supervisory signal to be initiated at the required remote annunciation.

Exhibit II.36 Elevator Zone—Elevator and Fire Alarm System Installed at Same Time

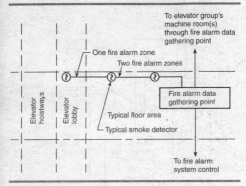

Source: NFPA 72®, National Fire Alarm Code®, NFPA, Quincy, MA, 1999 edition, Fig. A-3-9.3.7(a).

Editor's Note: Each group of elevators must terminate at the designated elevator controller within the group's elevator machine room(s) for fire fighter's service.

CONTROL FUNCTIONS

Exhibit II.37 Elevator Zone—Elevator Installed after Fire Alarm System

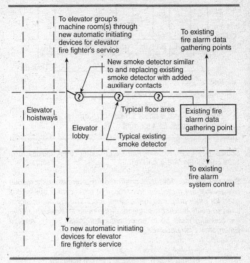

Source: NFPA 72®, National Fire Alarm Code®, NFPA, Quincy, MA, 1999 edition, Fig. A-3-9.3.7(b).

Exhibit II.38a Temporal Pattern Parameters

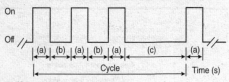

Key:

Phase (a) signal is on for 0.5 s ± 10%
Phase (b) signal is off for 0.5 s ± 10%
Phase (c) signal is off for 1.5 s ± 10% [(c) = (a) + 2(b)]
Total cycle lasts for 4 s ± 10%

Source: NFPA 72®, National Fire Alarm Code®, NFPA, Quincy, MA, 1999 edition, Fig. A-3-8.4.1.2(a).

Editor's Note: Audible alarm notification appliances must produce signals that are distinctive from those of other similar appliances used for other purposes in the same area. The fire alarm signal used to notify building occupants of the need to evacuate (leave the building) must be in accordance with ANSI 53.41, Audible Emergency Evacuation Signal.

Exhibit II.38b Temporal Pattern Imposed on Audible Notification Appliances That Emit a Continuous Signal While Energized

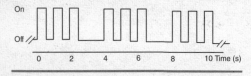

Source: NFPA 72®, National Fire Alarm Code®, NFPA, Quincy, MA, 1999 edition, Fig. A-3-8.4.1.2(b).

Exhibit II.38c Temporal Pattern Imposed on a Single Stroke Bell or Chime

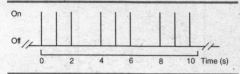

Source: NFPA 72®, National Fire Alarm Code®, NFPA, Quincy, MA, 1999 edition, Fig. A-8-1.2.3(c).

Exhibit II.39 Application of Survivability Requirements

	Applicability	Audible	Visible	Quantity (x)	Criteria
Total evacuation	N/A	N/A	N/A		
Partial evacuation	Yes	Yes	Yes		
Relocation of occupants	Yes	Yes	Yes		
One (1) NAC can serve x notification zones		Yes	Yes	1 only	
Failure of equipment or fault on one NAC cannot impair operation of any other NAC		Yes	Yes		Yes
A NAC and the circuit(s) for NAC operation must be protected from the FACU to the notificaiton zone point of entry		Yes	Yes		Yes

Editor's Note: Survivability (per 3-8.4.1.1 of NFPA 72) is applicable only to partial evacuation or relocation of occupants.

Exhibit II.40 Two Separated Class B NACs

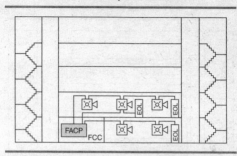

Editor's Note: This option can be used when the NACs are run directly from the FCC to the paging zone they serve.

Exhibit II.41 Class B NAC Riser in a 2-hr Rated Cable Assembly

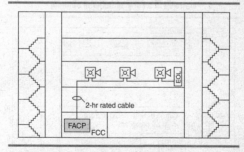

Editor's Note: This option can be used when it is not possible or desirable to use a 2-hr rated enclosure.

**Exhibit II.42 Class B NAC Riser
in a 2-hr Rated Shaft or Enclosure**

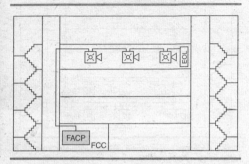

Editor's Note: This option can be used in lieu of a 2-hr rated cable.

**Exhibit II.43 Class B NAC Riser in a 2-hr Rated
Stairwell in a Fully Sprinklered Building**

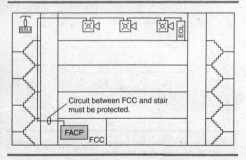

Editor's Note: Some local codes may prohibit this option. This option can be used in lieu of a 2-hr rated cable or enclosure.

Exhibit II.44 Class A NAC and a 2-hr Rated Cable Assembly

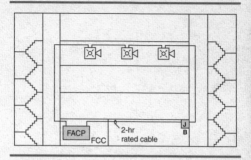

Editor's Note: Class A can be used; however, this option requires 2-hr protection of the circuits outside the paging zone served.

SECTION III

INITIATING DEVICES

The information in this section can be used to verify the correct location and spacing of detectors. It will be most useful to engineers, designers, contractors, installers, and inspectors. It includes the requirements for positioning detectors with regard to other detectors, structural and architectural features at the ceiling, and other building elements such as ductwork and lighting. Failure to properly locate and space detectors greatly increases the likelihood of delayed activation and of obstructed heat propagation and smoke migration patterns that reduce the effectiveness of the system. This section also provides information that can be used to verify distances between detectors that allow for their timely activation and result in the intended level of protection and response.

DETECTION CHECKLIST

Exhibit III.1 Detection Checklist

When selecting automatic detection, use the following
checklist to determine what type of detection is optimum for
the application.

DETECTION CHECKLIST

1. What type of detection coverage is required?
   a. Complete
   b. Partial
   c. Selective
   d. Supplementary

2. What are the hazards involved?
   a. Fuel loading
   b. Ignition source
   c. Slow vs. fast growth rate
   d. Smoldering vs. flaming rate
   e. Signatures
      1. Smoke (particle size)
      2. Heat
      3. Flame
      4. Gas
      5. Radiant energy

3. Who are the occupants?
   a. Mobile
   b. Assistance required
   c. Alert
   d. Mentally capable
   e. Familiar with exit plan
   f. Trained staff vs. general public

4. What is the environment?
   a. Large vs. small area
   b. Detectors suitable for the conditions

5. What are the goals of the system?

6. Is detector protection required? (Any mechanical guard
   must be listed for use with the detector.)

7. How large is the fire at the time of detection?

Exhibit III.1 Detection Checklist *Continued*

8. What suppression is available on the premises?
   a. Manual
   b. Automatic
9. How long before suppression is applied?
10. How long will it take for occupants to respond?
11. How long will evacuation take?

Exhibit III.2 Defining Characteristics for Joists

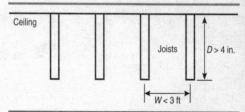

Exhibit III.3 Defining Characteristics for Beams

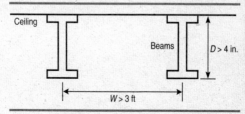

Exhibit III.4 Defining Characteristics for Girders

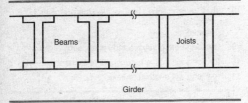

Girder

Editor's Note: A girder is a support for beams or joists that runs at right angles to the beams or joists. If the top of the girder is >4 in. from the ceiling, it is not a factor in detector location. If the top of the girder is <4 in. from the ceiling, the girder is to be considered a beam.

Exhibit III.5 Heat Detector Temperature Classification

Temperature Classification	Temperature Rating Range		Maximum Ceiling Temperature		Color Code
	°F	°C	°F	°C	
Low[a]	100–134	39–57	20 below	11 below	Uncolored
Ordinary	135–174	58–79	100	38	Uncolored
Intermediate	175–249	80–121	150	66	White
High	250–324	122–162	225	107	Blue
Extra high	325–399	163–204	300	149	Red
Very extra high	400–499	205–259	375	191	Green
Ultra high	500–575	260–302	475	246	Orange

[a]Intended only for installation in controlled ambient areas. Units shall be marked to indicate maximum ambient installation temperature.

Source: NFPA 72®, National Fire Alarm Code®, NFPA, Quincy, MA, 1999 edition; Table 2-2.1.1.1.

Editor's Note: The temperature rating of the detector must be at least 20°F (11°C) above the maximum expected temperature at the ceiling.

HEAT DETECTOR MOUNTING AND SPACING

Exhibit III.6 Proper Mounting for Heat Detectors

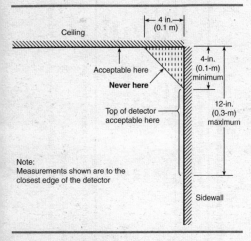

Source: NFPA 72®, National Fire Alarm Code®, NFPA, Quincy, MA, 1999 edition, Fig. A-2-2.1.

Exhibit III.7 Heat Detector Circle-of-Protection Coverage

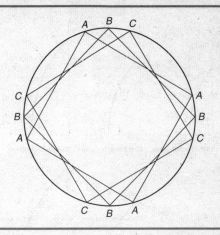

Source: NFPA 72®, National Fire Alarm Code®, NFPA Quincy, MA, 1999 edition, Fig. A-2-2.4.1(d).

Editor's Note: A detector with a 30 ft × 30 ft (9.1 m × 9.1 m) spacing is often used. A single detector covers an area that fits within the circle. For a rectangle, a single, properly located detector may be used provided the diagonal of the rectangle does not exceed the diameter of the circle. Detector coverage using circle-of-protection results in a circle whose radius is 0.7 times the selected spacing.

Exhibit III.8 Heat Detector Rectangular Area Layout Curves

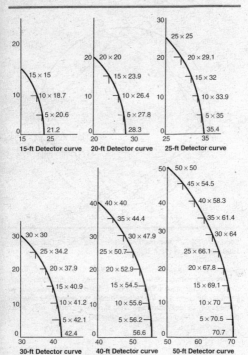

Source: NFPA 72®, National Fire Alarm Code®, NFPA, Quincy, MA, 1999 edition, Fig. A-2-2.4.1(e).

HEAT DETECTOR MOUNTING AND SPACING

Exhibit III.9 Heat Detector Spacing for Rectangular Areas

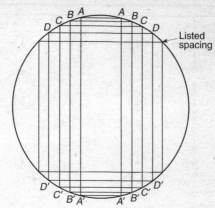

Rectangles
A = 10 ft × 41 ft = 410 ft²   (3.1 m × 12.5 m = 38.1 m²)
B = 15 ft × 39 ft = 585 ft²   (4.6 m × 11.9 m = 54.3 m²)
C = 20 ft × 37 ft = 740 ft²   (6.1 m × 11.3 m = 68.8 m²)
D = 25 ft × 34 ft = 850 ft²   (7.6 m × 10.4 m = 78.9 m²)
Listed spacing =
    30 ft × 30 ft = 900 ft²   (9.1 m × 9.1 m = 83.6 m²)

Source: NFPA 72®, National Fire Alarm Code®, NFPA, Quincy, MA, 1999 edition, Fig A-2-2.4.1(f).

Editor's Note: If the detector is not centered within the circle, the longer dimension should always be used in laying out detector coverage. As an example, a corridor 10 ft (3 m) wide and up to 82 ft (25 m) long can be covered by two 30 ft (9.1 m) detectors. Additional detectors are required when the rectangular dimensions are exceeded. It is often easier to properly place detectors by breaking down the area into multiple rectangles.

Exhibit III.10 Heat Detector Spacing

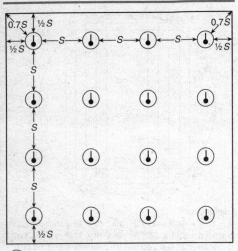

$\bigcirc\!\!\!|$ = Heat detector

S = Space between detectors

Source: NFPA 72®, National Fire Alarm Code®, NFPA, Quincy, MA, 1999 edition, Fig. A-2-2.4.1(a).

Editor's Note: If wall or partition extends up to within 18 in. (460 mm) of the ceiling, the space is considered a separate room.

Exhibit III.11 Heat Detector Spacing for Irregular Areas

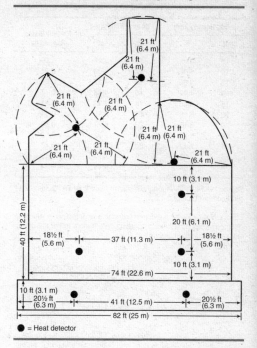

● = Heat detector

Source: NFPA 72®, National Fire Alarm Code®, NFPA, Quincy, MA, 1999 edition, Fig. A-2-2.4.1.2.

Editor's Note: No point on the ceiling can be more than 21 ft (6.4 m) away from a detector. Detector locations can be determined by striking arcs from the corner.

Exhibit III.12 Heat Detector Spacing Reduction Based on Ceiling Height

Ceiling Height Above		Up to and Including		Multiply Listed Spacing by
ft	m	ft	m	
0	0	10	3.05	1.00
10	3.05	12	3.66	0.91
12	3.66	14	4.27	0.84
14	4.27	16	4.88	0.77
16	4.88	18	5.49	0.71
18	5.49	20	6.10	0.64
20	6.10	22	6.71	0.58
22	6.71	24	7.32	0.52
24	7.32	26	7.93	0.46
26	7.93	28	8.54	0.40
28	8.54	30	9.14	0.34

Source: NFPA 72®, National Fire Alarm Code®, NFPA, Quincy, MA, 1999 edition, Table 2-2.4.5.1.

Editor's Note: Line-type electrical conductivity detectors in which resistance varies as a function of temperature must follow the manufacturer's recommendations.

**Exhibit III.13a Heat Detector Spacing
on Solid Joist Construction**

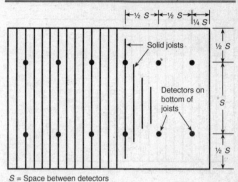

S = Space between detectors

*Source: NFPA 72®, National Fire Alarm Code®, NFPA, Quincy, MA,
1999 edition, Fig. A-2-2.4.5.1.*

**Exhibit III.13b Heat Detector Spacing
in Solid Joist Construction**

Joist Depth (D)	Joist Spacing (W)	Ceiling Height (H)	Mounting Criteria
<4 in.			Use smooth ceiling spacing
>4 in.	<3 ft		Right angle to joists, not to exceed 50% spacing
			Always install on bottom of joist

Exhibit III-14a Heat Detector Spacing on Beam Construction

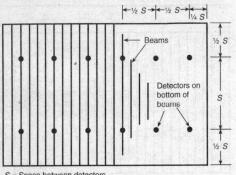

S = Space between detectors

Source: Modified from *NFPA 72®*, *National Fire Alarm Code®*, NFPA, Quincy, MA, 1999 edition, Fig. A-2-2.4.5.1.

Exhibit III.14b Heat Detector Spacing on Beam Construction

Beam Depth (D)	Beam Spacing (W)	Ceiling Height (H)	Mounting Criteria
<4 in.			Use smooth ceiling spacing
>4 in.			Right angles to beams, not to exceed 2/3 spacing
			Install using D/H criteria below
<12 in. AND	<8 ft		Right angle to beams, not to exceed 2/3 spacing
			Install using D/H criteria below
>18 in. AND	>8 ft		Treat each bay as a separate smooth ceiling space

Need to consider ratio of beam depth (D) to celing height (H)

If D/H > 0.10 AND W/H > 0.40	Treat each bay as a separate smooth ceiling space
If D/H < 0.10 OR W/H < 0.40	Right angles to beams not to exceed 50% spacing
	Install on bottom of beam

Exhibit III.15 Heat Detector Spacing on Peaked Ceilings

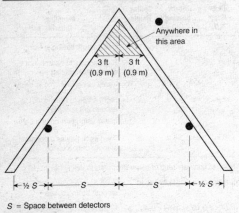

S = Space between detectors

● = Heat detector

Source: Modified from *NFPA 72®, National Fire Alarm Code®*, NFPA, Quincy, MA, 1999 edition, Fig. A-2-2.4.4.1.

Editor's Note: If roof slope is <30 degrees, space detectors using height at peak. If roof slope is >30 degrees, space detectors using average slope height.

Exhibit III.16 Heat Detector Spacing on Sloped Ceilings

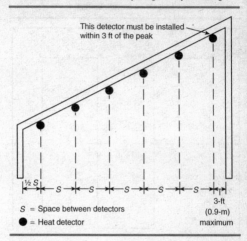

This detector must be installed within 3 ft of the peak

S = Space between detectors
● = Heat detector

½ S ← S → ← S → ← S → ← S → ← S →

3-ft
(0.9-m)
maximum

Source: Modified from *NFPA 72®*, *National Fire Alarm Code®*, NFPA, Quincy, MA, 1999 edition, Fig. A-2-2.4.4.2.

Editor's Note: If roof slope is <30 degrees, space detectors using height at peak. If roof slope is >30 degrees, space detectors using average slope height.

Exhibit III.17 Line-Type–Detector Spacing Layouts on Smooth Ceiling

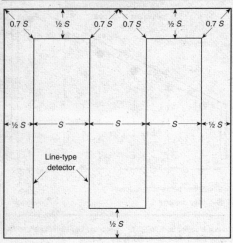

S = Space between detectors

Source: NFPA 72®, National Fire Alarm Code®, NFPA, Quincy, MA, 1999 edition, Fig. A-2-2.4.1(b).

Exhibit III.18 Conditions Affecting Smoke Detector Spacing and Location

Ceiling shape and surface
Ceiling height
Contents in the protected area
Fuel burning characteristics
Ventilation
Ambient environment

Editor's Note: Smoke detectors may not be appropriate in some environments. In this case, other types of detection must be used.

Exhibit III.19 Permitted Ambient Conditions for Smoke Detectors

Temperature between 32°F (0°C) and 100°F (38°C)
Relative humidity below 93%
Air velocity less than 300 ft/min (unless listed for such use)

Exhibit III.20 Detector Response Function vs. Particle Size

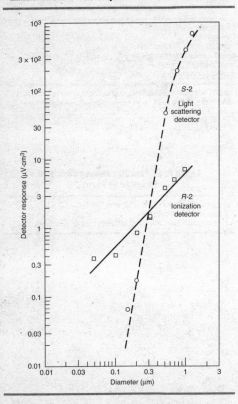

Source: Philip J. DiNenno et al., eds., *SFPE Handbook of Fire Protection Engineering,* 3rd ed., NFPA, Quincy, MA; Society of Fire Protection Engineers, Boston, 2002, Fig. 2-13.6.

Exhibit III.21 Smoke Particle Size from Burning Wood and Plastics

Type	d_{gm}, μm	d_{g2}, μm	σ	Combustion Conditions
Douglas fir	0.50–0.90	0.75–0.80	2.0	Pyrolysis
Douglas fir	0.43	0.47–0.52	2.4	Flaming
Polyvinylchloride	0.90–1.40	0.80–1.10	1.8	Pyrolysis
Polyvinylchloride	0.40	0.30–0.60	2.2	Flaming
Polyurethane (flexible)	0.80–1.80	0.80–1.00	1.8	Pyrolysis
Polyurethane (flexible)		0.50–0.70		Flaming
Polyurethane (rigid)	0.30–1.20	1.00	2.3	Pyrolysis
Polyurethane (rigid)	0.50	0.60	1.9	Flaming
Polystyrene		1.40		Pyrolysis
Polystyrene		1.30		Flaming
Polypropylene		1.60	1.9	Pyrolysis
Polypropylene		1.20	1.9	Flaming
Polymethylmethacrylate		0.60		Pyrolysis
Polymethylmethacrylate		1.20		Flaming
Cellulosic insulation	2.00–3.00		2.4	Smoldering

Source: Philip J. DiNenno et al., eds., *SFPE Handbook of Fire Protection Engineering*, 3rd ed., NFPA, Quincy, MA; Society of Fire Protection Engineers, Boston, 2002, Table 2-13.3.

Exhibit III.22 Environmental Conditions That Influence Smoke Detector Response

Detection Protection	Air Velocity >300 ft (>91.44 m)/min	Altitude >3000 ft (>914.4 m)	Humidity >93% RH	Temp. <32°F->100°F (<0°C->37.8°C)	Color of Smoke
Ion	X	X	X	X	O
Photo	O	O	X	X	X
Beam	O	O	X	X	O
Air sampling	O	O	X	X	O

X = Can affect detector response.

O = Generally does not affect detector response.

Source: NFPA 72®, National Fire Alarm Code®, NFPA, Quincy, MA, 1999 edition, Table A-2-3.6.1.1.

SMOKE DETECTOR RESPONSE CHARACTERISTICS

Exhibit III.23 Sources of Electrical and Mechanical Influences on Smoke Detectors

Electrical Noise and Transients	Airflow
Vibration or shock	Gusts
Radiation	Excessive velocity
Radio frequency	
Intense light	
Lightning	
Electrostatic discharge	
Power supply	

Source: NFPA 72®, National Fire Alarm Code®, NFPA, Quincy, MA, 1999 edition, Table A-2-3.6.1.2(b).

Exhibit III.24 Common Sources of Aerosols and Particulate Matter Moisture

Moisture
Humid outside air
Humidifiers
Live steam
Showers
Slop sink
Steam tables
Water spray

Combustion Products and Fumes
Chemical fumes
Cleaning fluids
Cooking equipment
Curing
Cutting, welding, and brazing
Dryers
Exhaust hoods
Fireplaces
Machining
Ovens
Paint spray

Atmospheric Contaminants
Corrosive atmospheres
Dust or lint
Excessive tobacco smoke
Heat treating
Linen and bedding handling
Pneumatic transport
Sawing, drilling, and grinding
Textile and agricultural
  processing

Engine Exhaust
Diesel trucks and locomotives
Engines not vented to the
  outside
Gasoline forklift trucks

**Heating Element with Abnormal
  Conditions**
Dust accumulations
Improper exhaust
Incomplete combustion

Source: NFPA 72®, National Fire Alarm Code®, NFPA, Quincy, MA, 1999 edition, Table A-2-3.6.1.2(a).

Exhibit III.25 Proper Mounting for Smoke Detectors

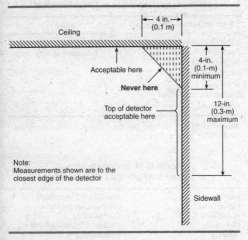

Note:
Measurements shown are to the closest edge of the detector

Source: NFPA 72®, National Fire Alarm Code®, NFPA, Quincy, MA, 1999 edition, Fig. A-2-2.2.1.

Exhibit III.26 Smoke Detector Circle-of-Protection Coverage

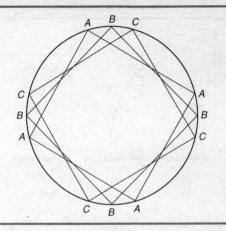

Source: NFPA 72®, National Fire Alarm Code®, NFPA, Quincy, MA, 1999 edition, Fig. A-2-2.4.1(d).

Editor's Note: A detector with a 30 ft × 30 ft (9.1 m × 9.1 m) spacing is often used. A single detector covers an area that fits within the circle. For a rectangle, a single, properly located detector may be used provided the diagonal of the rectangle does not exceed the diameter of the circle. Detector coverage using circle-of-protection results in a circle whose radius is 0.7 times the selected spacing.

**Exhibit III.27 Smoke Detector
Rectangular Area Layout Curves**

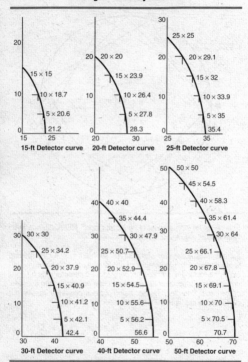

15-ft Detector curve

20-ft Detector curve

25-ft Detector curve

30-ft Detector curve

40-ft Detector curve

50-ft Detector curve

*Source: NFPA 72®, National Fire Alarm Code®, NFPA, Quincy, MA,
1999 edition, Fig. A-2-2.4.1(e).*

Exhibit III.28 Smoke Detector Spacing for Rectangular Areas

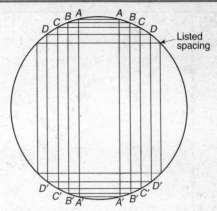

Rectangles
$A = 10$ ft $\times$ 41 ft = 410 ft^2 (3.1 m $\times$ 12.5 m = 38.1 m^2)
$B = 15$ ft $\times$ 39 ft = 585 ft^2 (4.6 m $\times$ 11.9 m = 54.3 m^2)
$C = 20$ ft $\times$ 37 ft = 740 ft^2 (6.1 m $\times$ 11.3 m = 68.8 m^2)
$D = 25$ ft $\times$ 34 ft = 850 ft^2 (7.6 m $\times$ 10.4 m = 78.9 m^2)
Listed spacing =
    30 ft $\times$ 30 ft = 900 ft^2 (9.1 m $\times$ 9.1 m = 83.6 m^2)

Source: NFPA 72®, National Fire Alarm Code®, NFPA, Quincy, MA, 1999 edition, Fig. A-2-2.4.1(f).

Editor's Note: If the detector is not centered within the circle, the longer dimension should always be used in laying out detector coverage. As an example, a corridor 10 ft (3 m) wide and up to 82 ft (25 m) long can be covered by two 30 ft (9.1 m) detectors. Additional detectors are required when the rectangular dimensions are exceeded. It is often easier to properly place detectors by breaking down the area into multiple rectangles.

Exhibit III.29 Smoke Detector Spacing

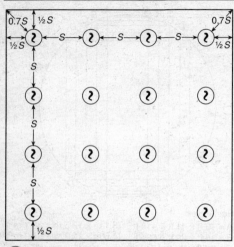

$?$ = Smoke detector

S = Selected spacing between detectors

Source: Modified from *NFPA 72®, National Fire Alarm Code®*, NFPA, Quincy, MA, 1999 edition, Fig. A-2-2.4.1(a).

Exhibit III.30 Smoke Detector Spacing for Irregular Areas

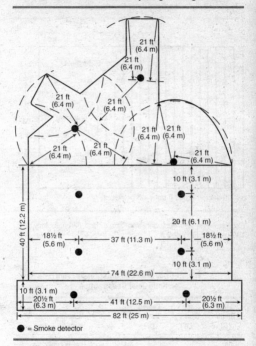

● = Smoke detector

Source: NFPA 72®, National Fire Alarm Code®, NFPA, Quincy, MA, 1999 edition, Fig. A-2-2.4.1.2.

Editor's Note: No point on the ceiling can be more than 21 ft (6.4 m) away from a detector. Detector locations can be determined by striking arcs from the corner.

SMOKE DETECTOR MOUNTING AND SPACING

Exhibit III.31a Smoke Detector Spacing
on Solid Joist Construction

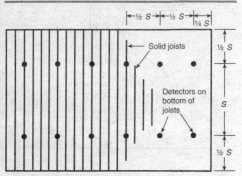

S = Selected spacing between detectors

Source: NFPA 72®, National Fire Alarm Code®, NFPA, Quincy, MA, 1999 edition, Fig A-2-2.4.5.1.

Exhibit III.31b Smoke Detector Spacing
on Solid Joist Construction

Joist Depth (D)	Joist Spacing (W)	Ceiling Height (H)	Mounting Criteria
<4 in.			Use smooth ceiling spacing
≤12 in.	AND	≤12 ft	Right angle to joists, not to exceed 50% spacing
			Always install on bottom of joist
>12 in.			Treat as beam

Exhibit III.32a Smoke Detector Spacing on Beam Construction

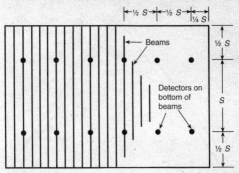

S = Selected spacing between detectors

Source: Modified from *NFPA 72®, National Fire Alarm Code®*, NFPA, Quincy, MA, 1999 edition, Fig A-2-2.4.5.1.

Exhibit III.32b Smoke Detector Spacing on Beam Construction

Beam Depth (D)	Beam Spacing (W)	Ceiling Height (H)	Mounting Criteria
<4 in.			Use smooth ceiling spacing
≤12 in. AND		≤12 ft	Right angle to beams, not to exceed 50% spacing
			Install on bottom of beam or on the ceiling
>12 in. OR		>12 ft	Treat each bay as a separate smooth ceiling space

Exhibit III.33 Smoke Detector Spacing on Peaked Ceilings

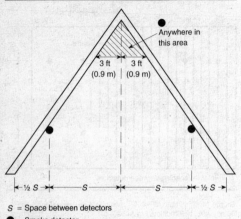

Anywhere in this area

3 ft (0.9 m) 3 ft (0.9 m)

½ S   S   S   ½ S

S = Space between detectors
● = Smoke detector

Source: Modified from *NFPA 72®, National Fire Alarm Code®*, NFPA, Quincy, MA, 1999 edition, Fig. A-2-2.4.4.1.

Exhibit III.34 Smoke Detector Spacing on Sloped Ceilings

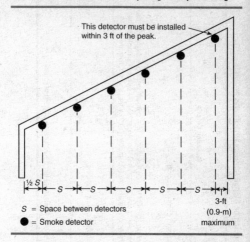

This detector must be installed within 3 ft of the peak.

½ S

S = Space between detectors

● = Smoke detector

3-ft (0.9-m) maximum

Source: Modified from *NFPA 72®, National Fire Alarm Code®*, NFPA, Quincy, MA, 1999 edition, Fig. A-2-2.4.4.2.

Exhibit III.35 Smoke Detector Spacing on Sloped Ceilings with Beams or Joists

Beam Depth (D)	Beam Spacing (W)	Ceiling Height (H)	Mounting Criteria
Beams parallel to (up) the slope			
>4 in.		Use average height, slope <10 degrees	Use flat beamed ceiling spacing
>4 in.		Use average height, slope >10 degrees	In the direction parallel to (up) the slope, use smooth ceiling spacing. In the direction perpendicular to (across) the slope, use 50 percent smooth ceiling spacing. Detectors at 50 percent spacing at low end are not required.
Beams perpendicular (across) the slope			
>4 in.		Use average height	Use flat beamed ceiling spacing
Joists			
>4 in.			Treat as a beam. Always install detectors on bottom of joists.

Exhibit III.36 Under Floor Mounting
Installations Permitted and Not Permitted

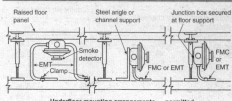

Underfloor mounting arrangements — permitted

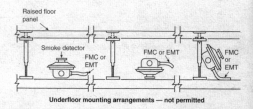

Underfloor mounting arrangements — not permitted

Source: NFPA 72®, National Fire Alarm Code®, NFPA, Quincy, MA, 1999 edition, Fig. A-2-3.4.3.2.

Editor's Note: Smoke detectors must be mounted only in an orientation for which they have been listed.

**Exhibit III.37a Smoke Detector
Spacing Based on High Air Movement**

Minimum per Air Change	Air Changes per Hour	Spacing per Detector	
		ft²	m²
1	60	125	11.61
2	30	250	23.23
3	20	375	34.84
4	15	500	46.45
5	12	625	58.06
6	10	750	69.68
7	8.6	875	81.29
8	7.5	900	83.61
9	6.7	900	83.61
10	6	900	83.61

*Source: NFPA 72®, National Fire Alarm Code®, NFPA, Quincy, MA,
1999 edition, Table 2-3.6.6.3.*

**Exhibit III.37b Smoke Detector Spacing
Based on High Air Movement**

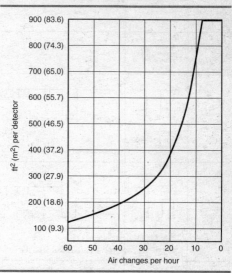

Source: *NFPA 72®, National Fire Alarm Code®*, NFPA, Quincy, MA,
1999 edition, Fig. 2-3.6.6.3.

Editor's Note: Not to be used for under-floor or above-ceiling spaces.

Exhibit III.38 Detector Location Requirements for Wall Sections

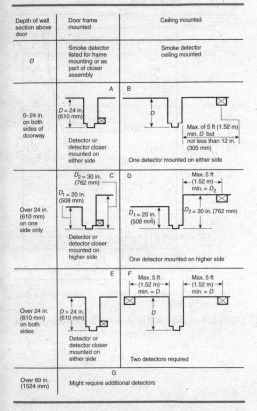

Source: NFPA 72®, National Fire Alarm Code®, NFPA, Quincy, MA, 1999 edition, Fig. 2-10.6.5.1.1.

Exhibit III.39 Detector Location Requirements for Single and Double Doors

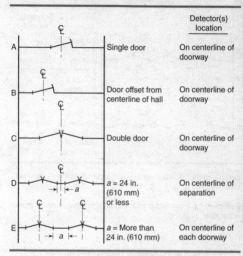

		Detector(s) location
A	Single door	On centerline of doorway
B	Door offset from centerline of hall	On centerline of doorway
C	Double door	On centerline of doorway
D	a = 24 in. (610 mm) or less	On centerline of separation
E	a = More than 24 in. (610 mm)	On centerline of each doorway

Source: NFPA 72®, National Fire Alarm Code®, NFPA, Quincy, MA, 1999 edition, Fig. 2-10.6.5.3.1.

Exhibit III.40 Detector Location
Requirements for Group Doorways

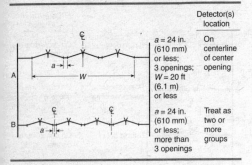

		Detector(s) location
A	a = 24 in. (610 mm) or less; 3 openings; W = 20 ft (6.1 m) or less	On centerline of center opening
B	a = 24 in. (610 mm) or less; more than 3 openings	Treat as two or more groups

Source: NFPA 72®, National Fire Alarm Code®, NFPA, Quincy, MA, 1999 edition, Fig. 2-10.6.5.3.2.

Editor's Note: Each group of the three doorways must be treated separately.

Exhibit III.41 Detector Location Requirements
for Group Doorways over 20 ft (6.1 m) in Width

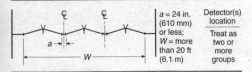

	Detector(s) location
a = 24 in. (610 mm) or less; W = more than 20 ft (6.1 m)	Treat as two or more groups

Source: NFPA 72®, National Fire Alarm Code®, NFPA, Quincy, MA, 1999 edition, Fig. 2-10.6.5.3.3.

Exhibit III.42a Location of Smoke Detectors in Return Air System Openings for Selective Operation of Equipment

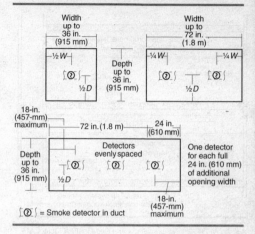

$\{Ⓓ\}$ = Smoke detector in duct

Source: NFPA 72®, National Fire Alarm Code®, NFPA, Quincy, MA, 1999 edition, Fig. A-2-10.4.2.2(a).

Exhibit III.42b Location of Smoke Detectors in Return Air System for Selective Operation of Equipment

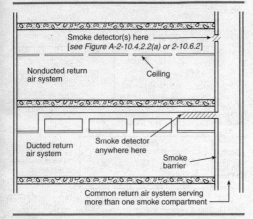

Source: NFPA 72®, National Fire Alarm Code®, NFPA, Quincy, MA, 1999 edition, Fig. A-2-10.4.2.2(b).

Exhibit III.42c Detector Location in a Duct That Passes through Smoke Compartments Not Served by the Duct

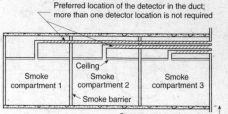

Source: NFPA 72®, National Fire Alarm Code®, NFPA, Quincy, MA, 1999 edition, Fig. A-2-10.4.2.2(c).

Exhibit III.43 Typical Single Zone Duct Detection System

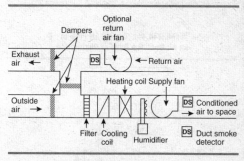

Source: Guide for Proper Use of Smoke Detectors in Duct Applications, NEMA, Washington, DC, 2001, Fig. 6-1.

Exhibit III.44a Typical Air Flow in an Air Duct at a Fan

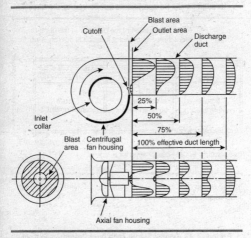

Source: Adapted with permission from AMCA Publication 201.

**Exhibit III.44b Typical Air Flow in
an Air Duct at a Bend or Obstruction**

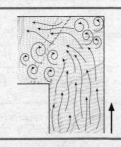

Exhibit III.45 Typical Duct Detector Placement

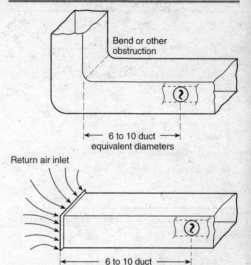

Source: NFPA 72®, *National Fire Alarm Code®*, NFPA, Quincy, MA, 1999 edition, Fig. A-2-10.5.2(b).

SMOKE DETECTORS FOR SMOKE CONTROL

Exhibit III.46 Cross Section of a Rectangular Air Duct Where the Air Flow Is Known

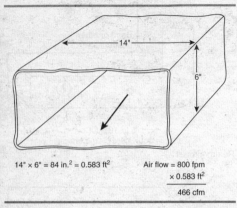

14" × 6" = 84 in.^{2} = 0.583 ft^2

$$
\begin{array}{r}
\text{Air flow = 800 fpm} \\
\times\ 0.583\ \text{ft}^2 \\
\hline
466\ \text{cfm}
\end{array}
$$

Editor's Note: Ensure duct smoke detector is listed for the duct's air velocity.

144   INITIATING DEVICES

Exhibit III.47 Cross Section of the Air Duct Depicting Air Flow Sampling Locations to Determine Air Volume

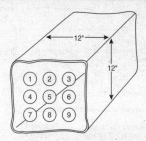

Cross sectional area of duct
12" × 12" = 144 in.^{2} = 1 ft^2

Reading #1	180 fpm
Reading #2	230 fpm
Reading #3	200 fpm
Reading #4	200 fpm
Reading #5	250 fpm
Reading #6	220 fpm
Reading #7	180 fpm
Reading #8	230 fpm
Reading #9	200 fpm

1890 ÷ 9 = 210 fpm
× 1 ft^2

210 cfm

Editor's Note: A velometer can be inserted into the duct at various cross section points to measure air velocity. The air velocity must be measured at various cross section points and the readings averaged to obtain an accurate velocity measurement. Ensure duct smoke detector is listed for the duct's air velocity.

Exhibit III.48 Cross Section of the Air Duct Depicting Air Flow Sampling Locations to Determine Air Volume

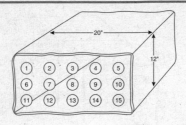

Cross sectional area of duct
20" × 12" = 240 in.^{2} = 1.66 ft^2

Reading #1	800 fpm
Reading #2	950 fpm
Reading #3	1,100 fpm
Reading #4	1,000 fpm
Reading #5	950 fpm
Reading #6	850 fpm
Reading #7	1,000 fpm
Reading #8	1,150 fpm
Reading #9	1,050 fpm
Reading #10	1,000 fpm
Reading #11	900 fpm
Reading #12	1,050 fpm
Reading #13	1,200 fpm
Reading #14	1,100 fpm
Reading #15	1,050 fpm

15,150 fpm
÷ 15

1,010 fpm
× 1.66 ft^2

1,676.6 cfm

Source: Modified from *NFPA 72®, National Fire Alarm Code®*, NFPA, Quincy, MA, 1999 edition, Fig. A-2-10.5.2(a).

Editor's Note: A velometer can be inserted into the duct at various cross section points to measure air velocity. The air velocity must be measured at various cross section points and the readings averaged to obtain an accurate velocity measurement. Ensure duct smoke detector is listed for the duct's air velocity.

Exhibit III.49 Cross Section of a Round Air Duct Where the Air Flow Is Known

12 in. diameter duct with 1,000 fpm velocity

$A = \pi r^2$

$A = 3.14159 \times 6^2$

$A = 3.14159 \times 36$

$A = 113 \text{ in.}^2$

$A = 0.78 \text{ ft}^2$

$V = 0.78 \times 1,000 = 780 \text{ cfm}$

Exhibit III.50 Equivalent Rectangular Duct Dimensions

Circular Duct Diameter in.	Length of One Side of Rectangular Duct (a), in.																			
	4	5	6	7	8	9	10	12	14	16	18	20	22	24	26	28	30	32	34	36
	Length Adjacent Side of Rectangular Duct (b), in.																			
5	5																			
5.5	6	5																		
6	8	6	6																	
6.5	9	7	6																	
7	11	8	7																	
7.5	13	10	8	7																
8	15	11	9	8																
8.5	17	13	10	9																
9	20	15	12	10	8															
9.5	22	17	13	11	9															
10	25	19	15	12	10	9														
10.5	29	21	16	14	12	10														
11	32	23	18	15	13	11	10													
11.5		26	20	17	14	12	11													
12		29	22	18	15	13	12													

12.5	32	24	20	17	15	13								
13	35	27	22	18	16	14	12							
13.5	38	29	24	20	17	15	13							
14		32	26	22	19	17	14							
14.5		35	28	24	20	18	15							
15		38	30	25	22	19	16							
16		45	36	30	25	22	18							
17			41	34	29	25	20	16						
18			47	39	33	29	23	17						
19			54	44	38	33	26	19	18					
20				50	43	37	29	21	19					
21				57	48	41	33	23	20					
22				64	54	46	36	26	23	20				
23					60	51	41	28	25	22				
24					66	57	44	31	27	24	22			
25						63	49	34	29	26	24			
26						69	54	37	32	28	26	24		
27						76	59	40	35	31	28	25		
28							64	43	38	33	30	27	26	
29							70	47	41	36	32	29	27	
30							76	51	44	39	35	31	29	28

Continued

Exhibit III.50 Equivalent Rectangular Duct Dimensions *Continued*

Length of One Side of Rectangular Duct (a), in.

Length Adjacent Side of Rectangular Duct (b), in.

Circular Duct Diameter, in.	4	5	6	7	8	9	10	12	14	16	18	20	22	24	26	28	30	32	34	36
31								82	66	55	47	41	37	34	31	29				
32								89	71	59	51	44	40	36	33	31				
33								96	76	64	54	48	42	38	35	33	30			
34									82	68	58	51	45	41	37	35	32			
35									88	73	62	54	48	44	40	37	34	32		
36									95	78	67	58	51	46	42	39	36	34		
37									101	83	71	62	55	49	45	41	38	36	34	
38									108	89	76	66	58	52	47	44	40	38	36	
39										95	80	70	62	55	50	46	43	40	38	36
40										101	85	74	65	58	53	49	45	42	40	38
41										107	91	78	69	62	56	51	47	44	42	40
42										114	96	83	73	65	59	54	50	46	44	42
43										120	102	88	77	69	62	57	53	49	46	43
44											107	93	81	73	66	60	55	51	48	45
45											113	98	86	76	69	63	58	54	50	47

46	120	103	90	80	72	66	61	56	53	49
47	126	108	95	84	76	69	64	59	55	52
48	133	114	100	89	80	73	67	62	58	54
49	140	120	105	93	84	76	70	65	60	56
50	147	126	110	98	88	80	73	68	63	59
51		132	115	102	92	83	76	71	66	61
52		139	121	107	96	87	80	74	69	64
53		145	127	112	100	91	83	77	71	67
54		152	133	117	105	95	87	80	74	70
55			139	123	110	99	91	84	78	72
56			145	128	114	104	95	87	81	75
57			151	134	119	108	98	91	84	78
58			158	139	124	112	102	94	87	81
59			165	145	130	117	107	98	91	85
60			172	151	135	122	111	102	94	88

Source: Copyright 2001, American Society of Heating, Refrigerating and Air-Conditioning Engineers, Inc. www.ashrae.org. Reprinted by permission from 2001 ASHRAE Handbook-Fundamentals.

Exhibit III.51 Typical Duct Velocities (ft/min)

	Velocity	Main Ducts		Branch Ducts		Outlets	
	Type	Supply	Return	Supply	Return	Supply	Return
Residential	Low	1000	800	600	600	500–750	500
Apartments, hotel bedrooms, hospital bedrooms	Low	1500	1300	1200	1000	500–750	500
Private offices, churches, director's rooms, libraries, schools	Low	1800	1400	1400	1200	500–1,000	600
Theaters, auditoriums	Low	1300	1100	1000	800		
General offices, up-scale restaurants, up-scale stores, banks	Medium	2000	1500	1600	1200	1,200–1,500	700
Other stores, cafeterias	Medium	2000	1500	1600	1200	1,500	800
Industrial	Medium	2500	1800	2200	1600		
Commercial institutions, public buildings	High	2500–3800	1400–1800	2000–3000	1200–1600		
Industrial	High	2500–4000	1800–2200	2200–3200	1500–1800		

SMOKE DETECTORS FOR SMOKE CONTROL

Exhibit III.52 Pendant-Mounted Air Duct Installation

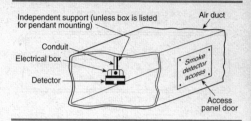

Source: NFPA 72®, National Fire Alarm Code®, NFPA, Quincy, MA, 1999 edition, Fig. A-2-10.5.2(a).

SMOKE DETECTORS FOR SMOKE CONTROL

Exhibit III.53 Inlet Tube Orientation

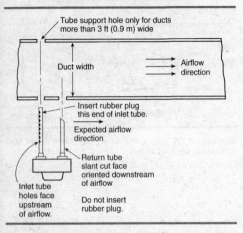

Source: NFPA 72®, National Fire Alarm Code®, NFPA, Quincy, MA, 1999 edition, Fig. A-2-10.5.2(c).

Editor's Note: Detectors installed in concealed locations more than 10 ft (3 m) above the finished floor require an accessible and clearly labelled remote alarm indicator to be installed except when the detector in alarm is indicated at the control unit.

SECTION IV

NOTIFICATION APPLIANCES

This section provides basic data to assist in the design, spacing, location, and installation of both audible and visible notification appliances. Tables and figures showing sound pressure levels, useful relationships, light criteria, and visible equivalency are included. This information, which is commonly used by design and installation personnel, is useful throughout the design and installation process.

Exhibit IV.1 Notification Appliance Mounting

Audible notification appliance	Top of appliance not less than 90 in. above finished floor and not less than 6 in. below ceiling
Visible notification appliance	Entire lens between 80 and 96 in. above finished floor
Combination audible-visible notification appliance	Entire lens between 80 and 96 in. above finished floor
Appliance application	Appliances for use in special environments (e.g., outdoors vs. indoors, high or low temperature, high humidity, dusty conditions, hazardous locations, areas subject ot tampering) must be listed for the intended application
Appliance protection	Any mechanical guard must be listed for use with the appliance

Exhibit IV.2 Audible Notification Appliance Requirements

	Private Mode	Public Mode[a]	Sleeping Area
Minimum appliance sound pressure level	45 dBA at 10 ft	75 dBA at 10 ft	—
Maximum appliance sound pressure level	120 dBA at minimum hearing distance from the appliance	120 dBA at minimum hearing distance from the appliance	—
Minimum system sound pressure level	The greater of 10 dBA above average ambient or 5 dBA above maximum sound pressure level lasting 60 s or longer measured 5 ft above the floor	The greater of 15 dBA above average ambient or 5 dBA above maximum sound pressure level lasting 60 s or longer measured 5 ft above the floor. An average ambient sound level greater than 105 dBA requires the use of a visible appliance(s).	—

Continued

Exhibit IV.2 Audible Notification Appliance Requirements *Continued*

	Private Mode	Public Mode[a]	Sleeping Area
Minimum sound pressure level	—	—	15 dBA above the average ambient sound level or 5 dBA above the maximum sound level having a duration of at least 60 s or a sound level of at least 70 dBA, whichever is greater, measured at the pillow level in the occupiable area. If any barrier, such as a door, curtain, or retractable partition, is located between the notification appliance and the pillow, the sound pressure level is measured with the barrier placed between the appliance and the pillow.

[a]Exceptions may apply.
Exception No. 1: Elevator cars
Exception No. 2: Restrooms
Exception No. 3: A system arranged to stop or reduce ambient noise if permitted by AHJ

Exhibit IV.3 Basic Audible System Design Criteria

Condition	Examples
Reflective acoustical surfaces	Tile, concrete, steel, gypsum wallboard, hardwood floors/walls, glass, and ice. Reflective surfaces do not attenuate sound pressure level as quickly as absorptive surfaces. Fewer appliances may be needed in a space with these surfaces.
Absorptive (soft) acoustical surfaces	Cloth, carpets, ceiling tiles, drapery, soft seating, and acoustical treatments. Soft surfaces attenuate sound pressure levels. More appliances may be needed.
Output of notification appliance	Generally 85 dBA at 10 ft. Check listing. Some appliances produce different sound pressure levels.
Distance of occupant from the source	Sound pressure level drops roughly 6 dBA when distance from source is doubled, depending on surfaces in the space. Consider additional appliances when occupant is far from source.
Direction of signal propagation	Signal must travel farther unless aimed at occupants.

Exhibit IV.4 Typical Audible Characteristics—Public Mode

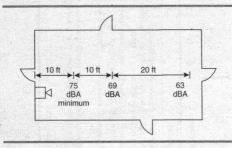

Editor's Note: Sound pressure level drops roughly 6 dBA when distance from source is doubled.

Exhibit IV.5 Typical Sound Propagation and Attenuation in a Space

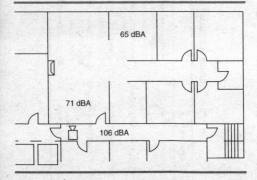

Exhibit IV.6 Attenuation through Typical Building Elements in 100–3150 Hz Frequency Range

Building element	Weight of partition (lb/sq ft)	Average attenuation (dB)
Walls and Partitions		
3.9 in. dense concrete with or without plaster	51.23	45
5.9 in 'no fines' concrete with 0.5 in. plaster on both faces	51.23	45
4.5 in. brickwork with 0.5 in. plaster on both faces	51.23	45
4.5 in. brickwork unplastered	39.96	42
11.8 in. lightweight concrete precast blocks with well-grouted joints	38.94	42
2.9 in. clinker blockwork with 0.5 in. plaster on both faces	23.57	40
1.9 in. dense concrete	24.59	40
1 in. plasterboard (2 layers) separated by timber studding < 2.9 in. and mineral fiber blanket		40
7.8 in. lightweight concrete precast blocks with well-grouted joints	25.00	40
5.9 in. lightweight concrete precast blocks with well-grouted joints	19.06	37
1.9 in. clinker blocks with 0.5 in. plaster on both faces		37
2.4 in. hollow clay blocks with 0.5 in. plaster on both faces	—	35
0.4 in. plasterboard (2 layers) separated by timber studding < 2.9 in. with 0.5 in. of plaster on both faces	—	35
0.02 in. plywood/hardboard (2 layers) separated by timber studding < 1.9 in. and 1.9 in. mineral fiber blanket	—	30

Continued

Exhibit IV.6 Attenuation through Typical Building Elements in 100–3150 Hz Frequency Range *Continued*

Building element	Weight of partition (lb/sq ft)	Average attenuation (db)
0.75 in. chipboard on a supporting frame	—	25
0.03 in. sheet steel	—	25
0.8 in. tongued and grooved softwood boards tightly clamped on a support frame	—	20
0.1 in hardboard (2 layers) separated by 1.7 in. polystyrene core	—	20
Doors		
Flush panel, hollow core, hung with one large air gap	1.84	14
Flush panel, hollow core hung with edge sealing	1.84	20
Solid hardwood, hung with edge sealing	5.74	26
Windows		
Single glass in heavy frame	3.07	24
Double glazed 0.3 in. panes in separate frames 1.9 in. cavity	12.71	34
Double glazed 0.25 in. panes in separate frames 3.9 in. cavity	22.95	38
Double glazed 0.25 in. and 0.3 in. panes in separate frames 7.8 in. cavity, absorbent blanket in reveals	44.06	58

Source: Philip J. DiNenno et al., eds., *SFPE Handbook of Fire Protection Engineering*, 3rd ed., NFPA, Quincy, MA: Society of Fire Protection Engineers, Boston, 2002. Modified from Table 3-1.17.

AUDIBLE NOTIFICATION APPLIANCES

Exhibit IV.7 Average Ambient Sound Level According to Location

Location	Average Ambient Sound Level (dBA)
Business occupancies	55
Educational occupancies	45
Industrial occupancies	80
Institutional occupancies	50
Mercantile occupancies	40
Piers and water-surrounded structures	40
Places of assembly	55
Residential occupancies	35
Storage occupancies	30
Thoroughfares, high density urban	70
Thoroughfares, medium density urban	55
Thoroughfares, rural and suburban	40
Tower occupancies	35
Underground structures and windowless buildings	40
Vehicles and vessels	50

Source: NFPA 72®, National Fire Alarm Code®, NFPA, Quincy, MA, 1999 edition, Table A-4-3.2.

**Exhibit IV.8 Audible Attenuation Design
Analysis of a 90 dBA Appliance**

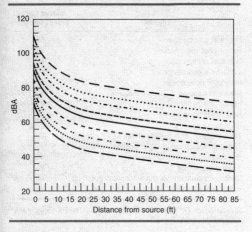

Exhibit IV.9 Two Methods of Visible Signaling: Direct vs. Indirect

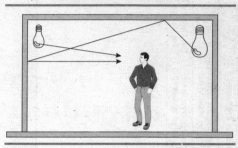

Exhibit IV.10 Visible Notification Appliance Synchronization Requirements

Number of Appliances	Synchronization Required
1	No
2	No
3 or more	No, if spaced a minimum of 55 ft (16.76 m) from each other in rooms 80 ft by 80 ft (24.4 m by 24.4 m) or greater
3 or more	Yes, if more than two visible notification appliances are in the same room or adjacent space within the same field of view

Exhibit IV.11 Room Spacing for Wall Mounted Visible Appliances

Maximum Room Size		Minimum Required Light Output (Effective Intensity) (cd)		
ft	m	One Light per Room	Two Lights per Room (Located on Opposite Wall)	Four Lights per Room; One Light per Wall
20 × 20	6.1 × 6.1	15	NA	NA
30 × 30	9.14 × 9.14	30	15	NA
40 × 40	12.2 × 12.2	60	30	15
50 × 50	15.2 × 15.2	95	60	30
60 × 60	18.3 × 18.3	135	95	30
70 × 70	21.3 × 21.3	185	95	60
80 × 80	24.4 × 24.4	240	135	60
90 × 90	27.4 × 27.4	305	185	95
100 × 100	30.5 × 30.5	375	240	95
110 × 110	33.5 × 33.5	455	240	135
120 × 120	36.6 × 36.6	540	305	135
130 × 130	39.6 × 39.6	635	375	185

NA: Not allowable.

Source: NFPA 72®, National Fire Alarm Code®, NFPA, Quincy, MA, 1999 edition, Table 4-4.1.1(a).

Exhibit IV.12 Room Spacing for a Single Wall Mounted Visible Appliance

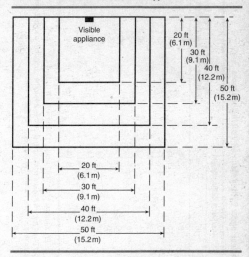

Source: NFPA 72®, National Fire Alarm Code®, NFPA, Quincy, MA, 1999 edition, Fig. 4-4.4.1.1.

Editor's Note: If a room is not square, the room size that allows it to be encompassed within the 20 ft × 20 ft, 30 ft × 30 ft, etc. square or allows it to be subdivided into multiple squares is used. Room spacing is based on locating the visible appliance at the halfway distance of the longest wall.

Exhibit IV.13 Room Spacing for Ceiling Mounted Visible Appliances

Maximum Room Size		Maximum Ceiling Height		Minimum Required Light Output (Effective Intensity); One Light (cd)
ft	m	ft	m	
20 × 20	6.1 × 6.1	10	3.05	15
30 × 30	9.14 × 9.14	10	3.05	30
40 × 40	12.2 × 12.2	10	3.05	60
50 × 50	15.2 × 15.2	10	3.05	95
20 × 20	6.1 × 6.1	20	6.1	30
30 × 30	9.14 × 9.14	20	6.1	45
40 × 40	12.2 × 12.2	20	6.1	80
50 × 50	15.2 × 15.2	20	6.1	115
20 × 20	6.1 × 6.1	30	9.14	55
30 × 30	9.14 × 9.14	30	9.14	75
40 × 40	12.2 × 12.2	30	9.14	115
50 × 50	15.2 × 15.2	30	9.14	150

Source: NFPA 72®, National Fire Alarm Code®, NFPA, Quincy, MA, 1999 edition, Table 4-4.1.1(b).

Editor's Note: For ceiling heights greater than 30 ft (9.14 m), visible notification appliances are to be suspended at or below 30 ft (9.14 m) or wall mounted. Ceiling mounted appliances are to be located at the center of the room.

Exhibit IV.14 Spacing of Wall Mounted
Visible Appliances in Rooms

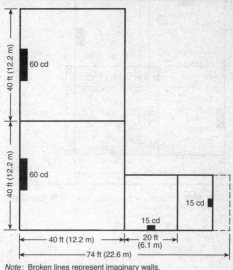

Note: Broken lines represent imaginary walls.

Source: NFPA 72®, National Fire Alarm Code®, NFPA, Quincy, MA, 1999 edition, Fig. A-4-4.4.1(b).

Exhibit IV.15a Room Spacing Allocation—Correct

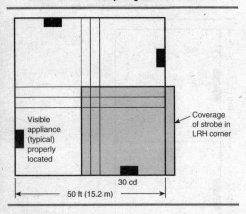

Visible appliance (typical) properly located

Coverage of strobe in LRH corner

30 cd

50 ft (15.2 m)

Source: NFPA 72®, National Fire Alarm Code®, NFPA, Quincy, MA, 1999 edition, Fig. A-4-4.4.1(c).

Exhibit IV.15b Room Spacing Allocation—Incorrect

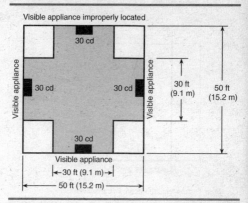

Source: NFPA 72®, *National Fire Alarm Code*®, NFPA, Quincy, MA, 1999 edition, Fig. A-4-4.4.1(d).

Exhibit IV.16a Irregular Area Spacing—Correct

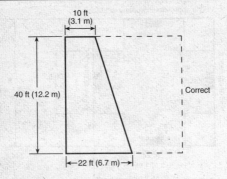

Note: Broken lines represent imaginary walls.

Source: NFPA 72®, National Fire Alarm Code®, NFPA, Quincy, MA, 1999 edition, Fig. A-4-4.4.1(a).

Exhibit IV.16b Irregular Area Spacing—Incorrect

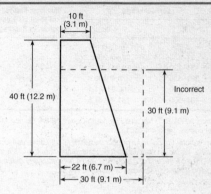

Note: Broken lines represent imaginary walls.

Source: NFPA 72®, National Fire Alarm Code®, NFPA, Quincy, MA, 1999 edition, Fig. A-4-4.4.1(a).

Exhibit IV.17 Corridor Spacing for Visible Appliances

Corridor Length		Minimum Number of 15-cd Visible Appliances Required
ft	m	
0–30	0–9.14	1
31–130	9.45–39.6	2
131–230	39.93–70	3
231–330	70.4–100.6	4
331–430	100.9–131.1	5
431–530	131.4–161.5	6

Source: NFPA 72®, National Fire Alarm Code®, NFPA, Quincy, MA, 1999 edition, Table 4-4.4.2.1.

Editor's Note: For corridors greater than 20 ft (6.1 m) wide, use room spacing. Visible appliances must be located not more than 15 ft (4.57 m) from the end of the corridor with a separation not greater than 100 ft (30.4 m) between appliances.

Exhibit IV.18 Corridor and Elevator Area Spacing Allocation

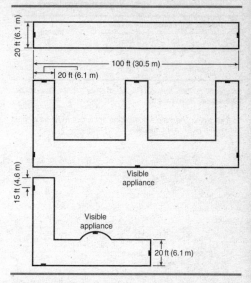

Source: *NFPA 72®, National Fire Alarm Code®*, NFPA, Quincy, MA, 1999 edition, Fig. A-4-4.2.

Exhibit IV.19 Effective Intensity Requirements for Sleeping Area Visible Notification Appliances

Distance from Ceiling to Top of Lens		Intensity (cd)
in.	mm	
≥24	610	110
<24	610	177

Source: NFPA 72®, National Fire Alarm Code®, NFPA, Quincy, MA, 1999 edition, Table 4-4.3.2.

Editor's Note: The visible notification appliance must be located within 16 ft (4.87 m) of the pillow.

Exhibit IV.20 Visible Equivalency

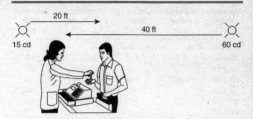

Editor's Note: Visible requirements are based on a specific source intensity spaced a certain distance apart or covering a specific area. Higher intensity light sources are needed as the distance between the source and viewer increases. The relationship between the light source and distance is found in Exhibit IX.13. An illuminance of 0.0375 lm/ft² from a visible appliance at any distance is considered equivalent.

Exhibit IV.21 Luminance Calculations

Maximum Room Size (ft)	One Light per Room (cd)	Square Room Size (d^2)	Illuminance $E = 0.375$ lm/ft^2 $E = I/d^2$	Luminance I (cd) $I = E \times d^2$
Per Exhibit IV.11		Calculated		
20 × 20	15	400	0.0375	15
25 × 25		625	0.0375	23
30 × 30	30	900	0.0375	34
35 × 35		1,225	0.0375	46
40 × 40	60	1,600	0.0375	60
45 × 45		2,025	0.0375	76
50 × 50	95	2,500	0.0375	94
55 × 55		3,025	0.0375	113
60 × 60	135	3,600	0.0375	135
65 × 65		4,225	0.0375	158
70 × 70	185	4,900	0.0375	184
75 × 75		5,625	0.0375	211
80 × 80	240	6,400	0.0375	240
85 × 85		7,225	0.0375	271
90 × 90	305	8,100	0.0375	304
95 × 95		9,025	0.0375	338
100 × 100	375	10,000	0.0375	375
105 × 105		11,025	0.0375	413
110 × 110	455	12,100	0.0375	454
115 × 115		13,225	0.0375	496
120 × 120	540	14,400	0.0375	540
125 × 125		15,625	0.0375	586
130 × 130	635	16,900	0.0375	634

VISIBLE EQUIVALENCY

Exhibit IV.22 Maximum Room Size Calculations

Maximum Room Size (ft)	One Light per Room (cd)	Illuminance $E = 0.375$ lm/ft² $E = I/d^2$	Luminance I (cd) $I = E \times d^2$	Square Room Size (d^2) $d^2 = I/E$
Per Exhibit IV.11		Calculated		
20 × 20	15	0.0375	15	400
25 × 25		0.0375	23	625
30 × 30	30	0.0375	34	900
35 × 35		0.0375	46	1,225
40 × 40	60	0.0375	60	1,600
44.72 × 44.72		0.0375	75	2,000
45 × 45		0.0375	76	2,025
50 × 50	95	0.0375	94	2,500
54.16 × 54.16		0.0375	110	2,933
55 × 55		0.0375	113	3,025
60 × 60	135	0.0375	135	3,600
65 × 65		0.0375	158	4,225
68.70 × 68.70		0.0375	177	4,720
70 × 70	185	0.0375	184	4,900
75 × 75		0.0375	211	5,625
80 × 80	240	0.0375	240	6,400
85 × 85		0.0375	271	7,225
90 × 90	305	0.0375	304	8,100
95 × 95		0.0375	338	9,025
100 × 100	375	0.0375	375	10,000
105 × 105		0.0375	413	11,025
110 × 110	455	0.0375	454	12,100
115 × 115		0.0375	496	13,225
120 × 120	540	0.0375	540	14,400
125 × 125		0.0375	586	15,625
130 × 130	635	0.0375	634	16,900

Editor's Note: Maximum room size is the square root of the square room size.

SECTION V

SUPERVISING STATIONS

This section is a collection of basic requirements for supervising station systems and related communications methods. It includes figures and tables showing the criteria for several supervising station types, transmission methods, loading, and other criteria. This information will be most useful to engineers, designers, inspectors, and other fire department personnel.

Exhibit V.1 Supervising Station Performance Criteria

System	Protected Premises	Central Station Service	Remote Supervising Station	Proprietary Supervising Station
Quality	All fire alarm systems	Supervising station service provided by a prime contractor. There is a subscriber (5-2.2.2 and 5-2.2.3).	Where central station service is neither required nor elected, properties under various ownership monitored by a remote supervising station (5-4.1)	Supervising station service monitoring contiguous or non-contiguous properties under one ownership and responsible to the owner of the protected property (5-3.2 et al.)
Listed	Equipment listed for the use intended (1-5.1.2)	Service listed or placarded as well as local equipment for intended use (5-1.2.1 and 5-1.2.2)	Equipment listed for use intended (1-5.1.2)	Equipment listed for use intended (1-5.1.2)
Design	According to code by experienced persons (1-5.1.3)	According to code by experienced persons (1-5.1.3)	According to code by experienced persons (1-5.1.3)	According to code by experienced persons (1-5.1.3)
Compatibility	Detector devices pulling power from initiating or signaling circuits listed for control panel (1-5.3)	Detector devices pulling power from initiating or signaling circuits listed for control panel (1-5.3)	Detector devices pulling power from initiating or signaling circuits listed for control panel (1-5.3)	Detector devices pulling power from initiating or signaling circuits listed for control panel (1-5.3)

Performance and limitations	85 percent and 110 percent of the nameplate rated input voltage, 32°F (0°C) and 120°F (49°C) ambient temperature, 85 percent relative humidity at 85°F (29.4°C)	85 percent and 110 percent of the nameplate rated input voltage, 32°F (0°C) and 120°F (49°C) ambient temperature, 85 percent relative humidity at 85°F (29.4°C)	85 percent and 110 percent of the nameplate rated input voltage, 32°F (0°C) and 120°F (49°C) ambient temperature, 85 percent relative humidity at 85°F (29.4°C)
Documentation	Authority having jurisdiction notified of new or changed specifications, wiring diagrams, battery calculations. Floor plans approval statement from contractor meets manufacturer's specifications and NFPA requirements. Record of completion (1-6.2.1 and 1-6.2.2). Results of evaluation required in 3-4.3.3.	Authority having jurisdiction notified of new or changed specifications, wiring diagrams, battery calculations. Floor plans approval statement from contractor meets manufacturer's specifications and NFPA requirements. Record of completion (1-6.2.1 and 1-6.2.2). Results of evaluation required in 3-4.3.3.	Authority having jurisdiction notified of new or changed specifications, wiring diagrams, battery calculations. Floor plans approval statement from contractor meets manufacturer's specifications and NFPA requirements. Record of completion (1-6.2.1 and 1-6.2.2). Results of evaluation required in 3-4.3.3.

Continued

Exhibit V.1 Supervising Station Performance Criteria *Continued*

System	Protected Premises	Central Station Service	Remote Supervising Station	Proprietary Supervising Station
Supervising station facilities	None	UL 827 compliant for both the supervising station and subsidiary station (5-1.2.1 and 5-1.2.2)	Access restricted, retransmit to public fire or governmental agency. Trouble and supervisory can be received elsewhere at a continuously attended station (5-4.3 et al.).	Fire resistive, detached building or cut-off room not near or exposed to hazards. Access restricted, NFPA 10, 26-hour emergency lighting (5-3.3 et al.)
Testing and maintenance	Chapter 7	Chapter 7	Chapter 7	Chapter 7
Runner service	No	Yes—alarm—arrive at the protected premises within 1 hr where equipment needs to be reset, guard signal—30 min, supervisory—1 hr, trouble—4 hr (5-2.6.1 et al.)	No	Yes—alarm—1 hr, guard alarm—30 min, supervisory—1 hr, trouble—1 hr (5-3.6.6 et al.)

Operations and management requirements	None	Central station provides all but test, maintenance, installation and runner service. Local alarm service provided by local alarm service prime contractor provides above and central station provides remainder (5-2.2.2).	None	See Qualify
Staff	None	Minimum of two persons on duty at supervising station. Operation and supervision primary task (5-2.5).	Sufficient to receive alarms. Other duties permitted per the authority having jurisdiction.	Two operators of which one may be the runner, when runner is not in attendance at station contact not to exceed 15 minutes. Primary duties are monitoring alarms and operations of station (5-3.5 et al).
Monitor signals	Control unit, fire command center and supervising station if sent to supervising station (1-5.4.3.2.2)	Control unit, fire command center and supervising station if sent to supervising station (1-5.4.3.2.2)	Control unit, fire command center and supervising station if sent to supervising station (1-5.4.3.2.2)	Control unit, fire command center and supervising station if sent to supervising station (1-5.4.3.2.2)

Continued

Exhibit V.1 Supervising Station Performance Criteria *Continued*

System	Protected Premises	Central Station Service	Remote Supervising Station	Proprietary Supervising Station
Retransmission	None—Local	Alarm—public fire service communications center, dispatch runner provide notice to subscriber and authority having jurisdiction if required. Supervisory, trouble, and guard service similar but not same (5-2.6.1.2–5-2.6.1.5).	(1) Alarm to fire department or governmental agency, if not available to another approved location. By dedicated phone circuit or one-way phone, private radio, or other methods acceptable. (2) Supervisory and trouble can go elsewhere (5-4.4.4).	Alarm—public fire department or plant brigade or other parties as required. Guard-protected premises, supervisory-designated person to find problem, others as required. Trouble-designated person to find problem and others as required (5-3.6.7.2–5-3.6.7.4).
Records	Current year and 1 yr after (1-6.3)	Complete records of all signals received must be retained for at least 1 yr. Reports provided of signals received to authority having jurisdiction in a form it finds acceptable (5-2.6.2).	At least 1 year (5-4.6.3).	Complete records of all signals received shall be retained for at least 1 yr. Reports provided of signals received to authority having jurisdiction in a form it finds acceptable (5-3.6.7).

Alarm type	Audible—fire zone. Audible and visual control unit and fire command center (1-5.7.1)	Audible—fire zone. Audible and visual control unit and fire command center (1-5.7.1)	Audible—fire zone. Audible and visual control unit and fire command center (1-5.7.1)	Audible—fire zone. Audible and visual control unit and fire command center (1-5.7.1)
Time/retransmit	20 s (1-5.4.2.2/None)	20 s (1-5.4.2.2) Immediate—public fire "maximum" 90 s (5-2.6.1.1)	20 s (1-5.4.2.2) Same/immediate—public fire or owner designate (5-4.6.2)	90 s (5-3.4.7) Immediate—public fire plant brigade or others (5-3.6.6.2)
Supervisory type	Audible—covered zone. Audible and visual at the control panel and fire control center (1-5.7.1)	Audible—covered zone. Audible and visual at the control panel and fire control center (1-5.7.1)	Audible—covered zone. Audible and visual at the control panel and fire control center (1-5.7.1)	Audible—covered zone. Audible and visual at the control panel and fire control center (1-5.7.1)
Time/retransmit	20 s (1-5.4.2.2/None)	20 s (1-5.4.2.2) Immediate—person designated by subscriber maximum 4 min (5-2.6.1.3)	20 s (1-5.4.2.2) Immediate—public fire or owner designate (5-4.6.2)	90 s (5-3.4.7) Where required immediate person designated (5-3.6.6.3)

Continued

Exhibit V.1 Supervising Station Performance Criteria *Continued*

System	Protected Premises	Central Station Service	Remote Supervising Station	Proprietary Supervising Station
Trouble type	Audible and visual at the control panel and fire control panel (1-5.7.1)	Audible and visual at the control panel and fire control panel (1-5.7.1)	Audible and visual at the control panel and fire control panel (1-5.7.1)	Audible and visual at the control panel and fire control panel (1-5.7.1)
Silence/reset	Can be silenced, can be intermittent every 10 s for 1/2 s, can have a switch secured. Re-sound every 24 hr if silenced, if re-stored and still silenced tone resounds (1-5.4.8).	Can be silenced, can be intermittent every 10 s for 1/2 s, can have a switch secured. Re-sound every 24 hr if silenced, if re-stored and still silenced tone resounds (1-5.4.8).	Can be silenced, can be intermittent every 10 s for 1/2 s, can have a switch secured. Re-sound every 24 hr if silenced, if re-stored and still silenced tone resounds (1-5.4.8).	Can be silenced, can be intermittent every 10 s for 1/2 s, can have a switch secured. Re-sound every 24 hr if silenced, if re-stored and still silenced tone resounds (1-5.4.8).
Time/retransmit	200 s (1-5.4.6/None)	200 s (1-5.4.6) Immediate—person designated by subscriber. Maximum 4 min (5-2.6.1.4)	200 s (1-5.4.6) Immediate—public fire or owner designate (5-4.6.2)	200 s (5-3.4.7) Where re-quired immediate person designated (5-3.6.6.4)

Primary source	Light and power service or engine-driven generator with trained operator (1-5.2.4). Low-power radio and DACT increased.	Same low-power radio and DACT increased	Same low-power radio and DACT increased	Same low-power radio and DACT increased
Capacity	Direct-current voltages not to exceed 350 V above earth ground (1-5.2.3)	Direct-current voltages not to exceed 350 V above earth ground (1-5.2.3)	Direct-current voltages not to exceed 350 V above earth ground (1-5.2.3)	Direct-current voltages not to exceed 350 V above earth ground (1-5.2.3)
Duration	Constant	Constant	Constant	Constant
Type	2-wire ac, 3-wire ac or dc with continuous unfused neutral conductor or polyphase ac with unfused neutral, dedicated branch circuits marked and over-current protection (1-5.2.4)	2-wire ac, 3-wire ac or dc with continuous unfused neutral conductor or polyphase ac with unfused neutral, dedicated branch circuits marked and over-current protection (1-5.2.4)	2-wire ac, 3-wire ac or dc with continuous unfused neutral conductor or polyphase ac with unfused neutral, dedicated branch circuits marked and over-current protection (1-5.2.4)	2-wire ac, 3-wire ac or dc with continuous unfused neutral conductor or polyphase ac with unfused neutral, dedicated branch circuits marked and over-current protection (1-5.2.4)

Continued

Exhibit V.1 Supervising Station Performance Criteria *Continued*

System	Protected Premises	Central Station Service	Remote Supervising Station	Proprietary Supervising Station
Secondary	(1) Storage battery. (2) Auto-starting engine generator and storage batteries with 4-hr capacity. (3) Multiple engine-driven generators, one auto starting and capable of largest generator being out of service. Does not cause loss of signal *(1-5.2.5)*	(1) Storage battery. (2) Auto-starting engine generator and storage batteries with 4-hr capacity. (3) Multiple engine-driven generators, one auto starting and capable of largest generator being out of service. Does not cause loss of signal *(1-5.2.5)*	(1) Storage battery. (2) Auto-starting engine generator and storage batteries with 4-hr capacity. (3) Multiple engine-driven generators, one auto starting and capable of largest generator being out of service. Does not cause loss of signal *(1-5.2.5)*	(1) Storage battery. (2) Auto-starting engine generator and storage batteries with 4-hr capacity. (3) Multiple engine-driven generators, one auto starting and capable of largest generator being out of service. Does not cause loss of signal *(1-5.2.5)*
Capacity	Direct-current voltages not to exceed 360 V above earth ground *(1-5.2.4)*	Direct-current voltages not to exceed 360 V above earth ground *(1-5.2.4)*	Direct-current voltages not to exceed 360 V above earth ground *(1-5.2.4)*	Direct-current voltages not to exceed 360 V above earth ground *(1-5.2.4)*

Duration	Automatic without signal loss for 30 s and then for 24 hr on protected premises (1-5.2.5)	Automatic without signal loss for 30 s and then for 24 hr on protected premises and central station (1-5.2.5)	Automatic without signal loss for 30 s and then for 60 hr on remote station (1-5.2.5)	Automatic without signal loss for 30 s and then for 24 hr on protected premises and proprietary station (1-5.2.5)
Switch	Automatic but see the different configurations for exception (1-5.2.6)	Automatic but see the different configurations for exception (1-5.2.6)	Automatic but see the different configurations for exception (1-5.2.6)	Automatic but see the different configurations for exception (1-5.2.6)
Type	Automatic (1-5.2.6)	Automatic (1-5.2.6)	Automatic (1-5.2.6)	Automatic (1-5.2.6)

Source: NFPA 72®, National Fire Alarm Code®, NFPA, Quincy, MA, 1999 edition, Table A-5-1.

Editor's Note: Numbers in parentheses refer to sections in NFPA 72, National Fire Alarm Code.

PERFORMANCE CRITERIA

Exhibit V.2 Central Station Service Elements

At Protected Premises	At Central Station
Installation of transmitters	Signal monitoring
Inspection	Retransmission
Testing	Record keeping
Maintenance	Reporting
Runner service	

Exhibit V.3 Central Station Indication of Compliance

Responsibility	Must be certificated or placarded Must be conspicuously posted
Prime contractor	Certificated • certified by organization that listed the Central Station • located on or within 3 ft of FACU Placarded • must meet requirements of organization that listed the Central Station • 20 sq in. or larger • located on or within 3 ft of FACU • identify Central Station by name and telephone number

Exhibit V.4 Subscriber Contracts for Central Station Services

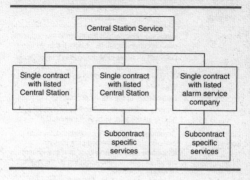

Exhibit V.5 Disposition of Signals for Central Station Service

Signal	Action Required
Alarm	1. Immediately retransmit the alarm to the public fire service communications center (within 90 s).
	2. Dispatch a runner or technician to the protected premises to arrive within 1 hr after receipt of a signal if equipment needs to be manually reset by the prime contractor.
	3. Immediately notify the subscriber.
	4. Provide notice to the subscriber or authority having jurisdiction, or both, if required.
Guard's tour supervisory	1. Communicate without reasonable delay with personnel at the protected premises by telephone, radio, calling back over the system circuit, or other means accepted by the authority having jurisdiction.
	2. Dispatch a runner to the protected premises to arrive within 30 min of the delinquency if communications with the guard cannot be promptly established.
	3. Report all delinquencies to the subscriber or authority having jurisdiction, or both if required.
Supervisory	1. Communication immediately with the person(s) designated by the subscriber (within 4 min).
	2. Dispatch a runner or maintenance person to arrive within 1 hr to investigate.
	3. Notify the fire department or law enforcement agency, or both, if required.
	4. Notify the authority having jurisdiction when sprinkler systems or other fire suppression systems or equipment has been wholly or partially out of service for 8 hr.
	5. When service has been restored, provide notice, if required, to the subscriber or the authority having jurisdiction, or both, as to the nature of the signal, the time of occurrence, and the restoration of service when equipment has been out of service for 8 hr or more.

**Exhibit V.5 Disposition of Signals
for Central Station Service *Continued***

Signal	Action Required
Trouble	1. Communicate immediately with persons designated by the subscriber (within 4 min).
	2. Dispatch personnel to arrive within 4 hr to initiate maintenance, if necessary.
	3. Provide notice, if required, to the subscriber or the authority having jurisdiction, or both, as to the nature of the interruption, the time of occurrence, and the restoration of service, when the interruption is more than 8 hr.

Exhibit V.6 Disposition of Signals for Proprietary Supervising Station Service

Signal	Action Required
Alarm	1. Immediately notify the fire department, the plant fire brigade, and such other parties as the authority having jurisdiction requires.
	2. Promptly dispatch a runner to the alarm location (travel time shall not exceed 1 hr).
	3. Restore the system as soon as possible after disposition of the cause of the alarm signal.
Guard's tour supervisory	1. Communicate immediately with the designated person(s) to ascertain the reason for the signal.
	2. Dispatch a runner or maintenance person (travel time not to exceed 1/2 hr for guard's tour, 1 hr for supervisory) to investigate, unless supervisory conditions are promptly restored.
	3. Notify the fire department.
	4. Notify the authority having jurisdiction when sprinkler systems are wholly or partially out of service for 8 hr or more.
	5. Provide written notice to the authority having jurisdiction as to the nature of the signal, time of occurrence, and restoration of service, when equipment has been out of service for 8 hr or more.
Trouble	1. Communicate immediately with the designated person(s) to ascertain reason for the signal.
	2. Dispatch a runner or maintenance person (travel time not to exceed 1 hr) to investigate.
	3. Notify the fire department.
	4. Notify the authority having jurisdiction when interruption of service exists for 4 hr or more.
	5. Provide written notice to the authority having jurisdiction as to the nature of the signal, time of occurrence, and restoration of service, when equipment has been out of service for 8 hr or more.

Exhibit V.7 Disposition of Signals
for Remote Supervising Station Service

Signal	Action Required
Alarm	If the remote supervising station is at a location other than the public fire service communications center, alarm signals shall be immediately retransmitted to the public fire service communications center.
Supervisory and trouble	Upon receipt of an alarm, supervisory, or trouble signal by the remote supervising station other than the public fire service communications center, the operator on duty shall be responsible for notifying the owner or the owner's designated representative immediately.

Exhibit V.8 Communications Methods for Supervising Stations

Criteria	5-5.3.1 Active Multiplex	5-5.3.2 Digital Alarm Communicator Systems	5-5.3.3 McCulloh Systems	5-5.3.4 Two-Way Radio Frequency (RF) Multiplex Systems
FCC approval when applicable	Yes	Yes	Yes	Yes
Conform to NFPA 70, National Electrical Code	Yes	Yes	Yes	Yes
Monitoring for integrity of the transmission and communications channel	Systems are periodically polled for end-to-end communications integrity	Both the premises unit and the system unit monitor for integrity in a manner approved for the means of transmission employed. A single signal received on each incoming DACR line once every 24 hr	Continuous dc supervision	Systems are periodically polled for end-to-end communications integrity
Annunciate, at the supervising station, the degradation and restoration of the transmission or communications channel	Within 200 s for Type III multiplex. Within 90 s for Type I and II multiplex.	Within 4 min using alternate phone line to report the trouble	Indicate automatically and operation under fault condition achieved either manually or automatically	Not exceed 90 s from the time of the actual failure

Redundant communication path where a portion of the transmission or communications channel cannot be monitored for integrity	Redundant path not required—supervising station always indicates a communications failure	Employ a combination of two separate transmission channels alternately tested at intervals not exceeding 24 hours	Redundant path not required—supervising station always indicates a communications failure. Exception is the use of nonmetallic channels that require two channels or an immediate transfer to a standby channel	Redundant path not required—supervising station always indicates a communications failure
Interval testing of the back-up path(s)	For Type I, 1-hr testing for dedicated lines and 24-hr testing for dial-up. No requirement for Type II and III.	When two phone lines are used, test alternately every 24 hr. Testing for other back-up technologies, see 5-5.3.2.1.6.2	No testing requirement	Back-up path not required
Annunciation of communication failure or ability to communicate at the protected premises	Not required—always annunciated at the supervising station that initiates corrective action	Indication of failure at premises due to line failure or failure to communicate after from 5 to 10 dialing attempts	Not required—always annunciated at the supervising station that initiates corrective action	Not required—always annunciated at the supervising station that initiates corrective action

Continued

Exhibit V.8 Communications Methods for Supervising Stations *Continued*

Criteria	5-5.3.1 Active Multiplex	5-5.3.2 Digital Alarm Communicator Systems	5-5.3.3 McCulloh Systems	5-5.3.4 Two-Way Radio Frequency (RF) Multiplex Systems
Time to restore signal receiving, processing, display, and recording equipment	Where duplicate equipment not provided, spare hardware required so a repair can be effected within 30 min	Spare digital alarm communicator receivers required for switchover to back-up receiver in 30 s. One back-up system unit for every 5 system units.	Where duplicate equipment not provided, spare hardware required so a repair can be effected within 30 min	Where duplicate equipment not provided, spare hardware required so a repair can be effected within 30 min
Loading capacities for system units and transmission and communications channels	512 buildings and premises on one system unit. Unlimited if you can switch to duplicate system unit within 30 s. Loading capacity for transmission and communications channels (trunks) is listed in Table 5-5.3.1.4.	See Table 5-5.3.2.2.3 for the maximum number of transmitters on a hunt group in a system unit	Alarm and sprinkler supervisory limited to 25 plants and 250 code wheels on one circuit. 60 scheduled guard reports per hour	512 buildings and premises on a system unit with no back-up. Unlimited if you can switch to a back-up in 30 s.

End-to-end communication time for an alarm	90 s from initiation until it is recorded	Off-hook to on-hook not to exceed 90 s per attempt. 10 attempts maximum. 900 s maximum for all attempts.	90 s from initiation until it is recorded
Record and display rate of subsequent alarms at supervising station	Not slower than one every 10 additional seconds	Not addressed	When any number of subsequent alarms come in, record at a rate not slower than one every additional 10 s
Signal error detection and correction	Not addressed	Signal repetition, digital parity check, or some equivalent means of signal verification must be used	Not addressed
Path sequence priority	Not addressed	The first transmission attempt uses the primary channel	Not addressed

Continued

Exhibit V.8 Communications Methods for Supervising Stations *Continued*

Criteria	5-5.3.1 Active Multiplex	5-5.3.2 Digital Alarm Communicator Systems	5-5.3.3 McCulloh Systems	5-5.3.4 Two-Way Radio Frequency (RF) Multiplex Systems
Carrier diversity	None required	Where long distance service (including WATS) is used, the second telephone number shall be provided by a different long distance service provider where there are multiple providers	Not addressed	Not addressed
Throughput probability	Not addressed	Demonstrate 90% probability of a system unit immediately answering a call or follow the loading Table 5-5.3.2.2.3. One-way radio back-up demonstrates 90% probability of transmission.	Not addressed	Not addressed

Unique premises identifier	Yes	Yes	Yes	Yes
Unique flaws	None addressed	If call forwarding is used to communicate to the supervising station, verify the integrity of this feature every 4 hr	None addressed	None addressed
Signal priority	Fire alarm, supervisory, and trouble signals shall take precedence, in that respective order of priority, over all other signals (except life threatening signals over supervisory and trouble)	Chapter 1 on fundamentals requires that alarm signals take priority over supervisory signals unless there is sufficient repetition of the alarm signal to prevent the loss of an alarm signal	Chapter 1 on fundamentals requires that alarm signals take priority over supervisory signals unless there is sufficient repetition of the alarm signal to prevent the loss of an alarm signal	Chapter 1 on fundamentals requires that alarm signals take priority over supervisory signals unless there is sufficient repetition of the alarm signal to prevent the loss of an alarm signal
Sharing communications equipment on premises	Not addressed	Disconnect outgoing or incoming telephone call and prevent its use for outgoing telephone calls until signal transmission has been completed	Not addressed	Not addressed

Continued

Exhibit V.8 Communications Methods for Supervising Stations *Continued*

Criteria	5-5.3.5 One-Way Private Radio Alarm Systems	5-5.3.6 Directly Connected Noncoded Systems	5-5.3.7 Private Microwave Radio Systems	5-5.4 Other Transmission Technologies
FCC approval when applicable	Yes	Yes	Yes	Yes
Conform to NFPA 70, *National Electrical Code*	Yes	Yes	Yes	Yes
Monitoring for integrity of the transmission and communications channel	Test signal from every transmitter once every 24 hr	Continuous dc supervision	Used as a portion of another type of transmission technology. End-to-end integrity monitored by the main transmission technology. Microwave portion is continuously monitored	Monitor for integrity or provide back-up channel tested as below
Annunciate, at the supervising station, the degradation and restoration of the transmission or communications channel	Only monitor the quality of signal received and indicate if the signal falls below minimum signal quality specified in code	Presented in a form to expedite prompt operator interpretation	Presented in a form to expedite prompt operator interpretation	Within 5 min (may use a second separate path to report failure)

Redundant communication path where a portion of the transmission or communications channel cannot be monitored for integrity	Minimum of two independent RF paths must be simultaneously employed	None required	Dual transmitters required if more than 5 buildings or premises or 50 initiating devices circuits	Provide a redundant path if communication failure not annunciated at supervising station
Interval testing of the back-up path(s)	No requirement because the quality of the signal is continuously monitored	Back-up path not required	Dual transmitters shall be operated on time ratio of 2:1 within each 24 hr	If back-up path required, test path once every 24 hr or on alternating channels testing each channel every 48 hr
Annunciation of communication failure or ability to communicate at the protected premises	Monitor the interconnection of the premises unit elements of transmitting equipment and indicate a failure at the premises or transmit a trouble signal to the supervising station	None required	None required	Systems where the transmitter at the local premises unit detects a communication failure before the supervising station, the premises unit will annunciate the failure within 5 min of detecting the failure

Continued

Exhibit V.8 Communications Methods for Supervising Stations *Continued*

Criteria	5-5.3.5 One-Way Private Radio Alarm Systems	5-5.3.6 Directly Connected Noncoded Systems	5-5.3.7 Private Microwave Radio Systems	5-5.4 Other Transmission Technologies
Time to restore signal receiving, processing, display, and recording equipment	Where duplicate equipment not provided, spare hardware required so a repair can be effected within 30 min	Where duplicate equipment not provided, spare hardware required so a repair can be effected within 30 min	Where duplicate equipment not provided, spare hardware required so a repair can be effected within 30 min	Where duplicate equipment not provided, spare hardware required so a repair can be effected within 30 min. Complete set of critical spare parts on a 1 to 5 ratio of parts to system units or a duplicate functionally equivalent system unit for every 5 system units

Loading capacities for system transmission units and transmission and communications channels	512 buildings and premises on a system unit with no back-up. Unlimited if you can switch to a back-up in 30 s.	A single circuit must not serve more than one plant.	Up to 5 buildings or premises or 50 initiating device circuits on one transmitter. Unlimited if dual transmitters are used with automatic switchover or manual switchover in 30 s	512 independent fire alarm systems on a system unit with no back-up. Unlimited if you can switch to a back-up in 30 s. The system shall be designed such that a failure of a transmission channel serving a system unit shall not result in the loss in the ability to monitor more than 3000 transmitters
End-to-end communication time for an alarm	90% probability to receive an alarm in 90 s, 99% probability in 180 s, 99.999% probability in 450 s	Not addressed	Not addressed	90 s from initiation of alarm until displayed to the operator and recorded on a media from which the information can be retrieved

Continued

Exhibit V.8 Communications Methods for Supervising Stations *Continued*

Criteria	5-5.3.5 One-Way Private Radio Alarm Systems	5-5.3.6 Directly Connected Noncoded Systems	5-5.3.7 Private Microwave Radio Systems	5-5.4 Other Transmission Technologies
Record and display rate of subsequent alarms at supervising station	When any number of subsequent alarms come in, record at a rate not slower than one every additional 10 s	Not addressed	Not addressed	No slower than one every 10 additional seconds
Signal error detection and correction	Not addressed	Not applicable	Not addressed	Signal repetition, parity check, or some equivalent means of error detection and correction shall be used
Path sequence priority	Not addressed	Not addressed	Not addressed	No need for prioritization of paths. The requirement is that both paths are equivalent
Carrier diversity	Not addressed	Not addressed	Not addressed	When a redundant path is required, the alternate path shall be provided by a public communication service provider different from the primary path where available

Throughput probability	90% probability to receive an alarm in 90 s, 99% probability in 180 s, 99.999% in probability 450 s	Not addressed	Not addressed	When the supervising station does not regularly communicate with the transmitter at least once every 200 s, then the throughput probability of the alarm transmission must be at least 90% in 90 s, 99% in 180 s, 99.999% in 450 s
Unique premises identifier	Yes	Yes	Yes	If a transmitter shares a transmission or communication channel with other transmitters, it shall have a unique transmitter identifier

Continued

Exhibit V.8 Communications Methods for Supervising Stations *Continued*

Criteria	5-5.3.5 One-Way Private Radio Alarm Systems	5-5.3.6 Directly Connected Noncoded Systems	5-5.3.7 Private Microwave Radio Systems	5-5.4 Other Transmission Technologies
Unique flaws	None addressed	None addressed	None addressed	From time to time, there may be unique flaws in a communication system. Unique requirements shall be written for these unique flaws
Signal priority	Chapter 1 on fundamentals requires that alarm signals take priority over supervisory signals unless there is sufficient repetition of the alarm signal to prevent the loss of an alarm signal	Chapter 1 on fundamentals requires that alarm signals take priority over supervisory signals unless there is sufficient repetition of the alarm signal to prevent the loss of an alarm signal	Chapter 1 on fundamentals requires that alarm signals take priority over supervisory signals unless there is sufficient repetition of the alarm signal to prevent the loss of an alarm signal	If the communication methodology is shared with any other usage, all fire alarm transmissions shall preempt and take precedence over any other usage. Fire alarm signals take precedent over supervisory signals

Sharing communications equipment on premises	Not addressed	Not addressed	Not addressed	If the transmitter is sharing on-premises communications equipment, the shared equipment shall be listed for the purpose (otherwise the transmitter must be installed ahead of the unlisted equipment)

Source: NFPA 72®, National Fire Alarm Code®, NFPA, Quincy, MA, 1999 edition. Table A-5-5.1.

Editor's Note: Cross-references are to NFPA 72, National Fire Alarm Code.

COMMUNICATIONS METHODS

Exhibit V.9 Communications Media for a DACT

Primary Method	Backup Method
Telephone line (number)	Telephone line (number)
Telephone line (number)	Cellular telephone connection
Telephone line (number)	One-way radio system
One telephone line (number) equipped with a derived local channel	None
Telephone line (number)	One-way private radio alarm system
Telephone line (number)	Private microwave radio system
Telephone line (number)	Two-way RF multiplex system
A single integrated services digital network (ISDN) telephone line using a terminal adapter specifically listed for supervising station fire alarm service, where the path between the transmitter and the switched telephone network serving central office is monitored for integrity so that the occurrence of an adverse condition in the path shall be annunciated at the supervising station within 200 s	None

Editor's Note: A DACT requires two means of communication, one of which must be a telephone line. Telephone lines must be loop start and cannot be ground start lines, such as those found on a Private Branch Exchange (PBX).

Exhibit V.10 Loading Capacities for Hunt Groups

System Loading at the Supervising Station	Number of Lines in Hunt Group				
	1	2	3	4	5–8
With DACR lines processed in parallel					
Number of initiating circuits	NA	5,000	10,000	20,000	20,000
Number of DACTs[a]	NA	500	1,500	3,000	3,000
With DACR lines processed serially (put on hold, then answered one at a time)					
Number of initiating circuits	NA	3,000	5,000	6,000	6,000
Number of DACTs[a]	NA	300	800	1,000	1,000

NA: Not allowed.

[a]Based on an average distribution of calls and an average connected time of 30 s for a message. The loading figures in the table shall presume that the lines are in a hunt group, that is, DACT is able to access any line not in use. A single-line DACR shall not be allowed (NA) for any of the configurations shown.

Source: Modified from NFPA 72®, National Fire Alarm Code®, NFPA, Quincy, MA, 1999 edition, Table 5-5.3.2.2.2.3.

Exhibit V.11 Loading Capacities for Active Multiplex Systems

Trunks	System Type		
	Type 1	Type 2	Type 3
Maximum number of fire alarm service initiating device circuits per primary trunk facility	5,120	1,280	256
Maximum number of leg facilities for fire alarm service per primary trunk facility	512	128	64
Maximum number of leg facilities for all types of fire alarm service per secondary trunk facility[a]	128	128	128
Maximum number of all types of initiating device circuits per primary trunk facility in any combination[a]	10,240	2,560	512
Maximum number of leg facilities for all types of fire alarm service per primary trunk facility in any combination[a]	1,024	256	128

System Units at the Supervising Station

Maximum number of all types of initiating device circuits per system unit[a]	10,240[b]	10,240[b]	10,240[b]
Maximum number of fire protecting buildings and premises per system unit	512[b]	512[b]	512[b]
Maximum number of fire alarm service initiating device circuits per system unit	5,120[b]	5,120[b]	5,120[b]
Systems Emitting from Subsidiary Station[c]	—	—	—

[a] Includes every initiating device circuit (for example, waterflow, fire alarm, supervisory, guard, burglary, hold-up).
[b] Paragraph 5-5.3.1.5 of NFPA 72 shall apply.
[c] Same as system units at the supervising station.

Source: NFPA 72®, National Fire Alarm Code®, NFPA, Quincy, MA, 1999 edition, Table 5-5.3.1.4.

COMMUNICATIONS METHODS

Exhibit V.12 Loading Capacities for Two-Way RF Multiplex Systems

Trunks	System Type	
	Type 4	Type 5
Maximum number of fire alarm service initiating device circuits per primary trunk facility	5,120	1,280
Maximum number of leg facilities for fire alarm service per primary trunk facility	512	128
Maximum number of leg facilities for all types of fire alarm service per secondary trunk facility[a]	128	128
Maximum number of all types of initiating device circuits per primary trunk facility in any combination	10,240	2,560
Maximum number of leg facilities for types of fire alarm service per primary trunk facility in any combination[a]	1,024	256
System Units at the Supervising Station		
Maximum number of all types of initiating device circuits per system unit[a]	10,240[b]	10,240[b]
Maximum number of fire protected buildings and premises per system unit	512[b]	512[b]
Maximum number of fire alarm service initiating device circuits per system	5,120[b]	5,120[b]
Systems Emitting from Subsidiary Station[c]	—	—

[a]Includes every initiating device circuit (for example, waterflow, fire alarm supervisory, guard, burglary, hold-up).

[b]Paragraph 5-5.3.4.5.2 of NFPA 72 shall apply.

[c]Same as system units at the supervising station.

Source: NFPA 72®, National Fire Alarm Code®, NFPA, Quincy, MA, 1999 edition, Table 5-5.3.4.5.1.

Exhibit V.13 Loading Capacities
of One-Way Radio Alarm Systems

	System Type	
Radio Alarm Repeater Station Receiver (RARSR)	Type 6	Type 7
Maximum number of fire alarm service initiating device circuits per RARSR	5,120	5,120
Maximum number of RATs for fire	512	512
Maximum number of all types of initiating device circuits per RARSR in any combination[a]	10,240	10,240
Maximum number of RATs for all types of fire alarm service per RARSR in any combination[a,c]	1,024	1,024
System Units at the Supervising Station		
Maximum number of all types of initiating device circuits per system unit[a]	10,240[b]	10,240[b]
Maximum number of fire-protected buildings and premises per system unit.	512[b]	512[b]
Maximum number of fire alarm service initiating device circuits per system unit	5,120[b]	5,120[b]

Notes:

1. Each guard tour transmitter shall reduce the allowable RATs by 15.

2. Each two-way protected premises radio transmitter shall reduce the allowable RATs by two.

[a]Includes every initiating device circuit (for example, waterflow, fire alarm, supervisory, guard, burglary, hold-up).

[b]Paragraph 5-5.3.5.7 of *NFPA 72* shall apply.

[c]Each supervised BA (open/close) or each suppressed guard tour transmitter shall reduce the allowable RATs by five.

Source: NFPA 72®, National Fire Alarm Code®, NFPA, Quincy, MA, 1999 edition, Table 5-5.3.5.6.

SECTION VI

INSPECTION, TESTING, AND MAINTENANCE

This section summarizes the methods, equipment, procedures, tables, and formulas necessary for proper fire alarm system inspection, testing, and maintenance. It describes the methods and procedures for conducting the tests. It also provides data relative to the type of tests required and schedules as to the frequency of each test.

The checklists included in this section can be used to ensure that service personnel conduct fire alarm tests effectively. The test methods table (Exhibit VI.2) contains a list of documentation, materials, tools, and procedures that are necessary for a successful test. The testing frequencies table (Exhibit VI.3) and visual inspection frequencies table (Exhibit VI.4) help service personnel minimize disruptions to occupants during testing and restore the system efficiently.

Exhibit VI.1 Test Checklist

Editor's Note: The following checklists can be used to ensure that service personnel conduct fire alarm tests efficiently. The first contains a list of documentation, materials, tools, and procedures that are necessary for a successful test. The second helps service personnel minimize disruptions to occupants during testing and restore the system efficiently.

BEFORE TESTING

Are the service personnel qualified and experienced in fire alarm system inspection, testing and maintenance?

Is all system documentation present?

- ❑ System certificate
- ❑ Floor plans
- ❑ Riser diagrams
- ❑ Specifications
- ❑ Manuals
- ❑ Wiring diagrams
- ❑ Operation matrices
- ❑ Record of completion
- ❑ System alterations
- ❑ Previous test records

Have the occupants been notified of the testing?

- ❑ E-mail
- ❑ Signs
- ❑ Bulletin boards
- ❑ Memorandums
- ❑ Announcements

Has the supervising station been notified of the testing?

Has the fire department been notified of the testing?

Are other trades present?

- ❑ Elevator service company
- ❑ Sprinkler service company

TEST CHECKLIST

Exhibit VI.1 Test Checklist *Continued*

- ❏ Special hazards systems service company
- ❏ HVAC service company

Do you have a contingency plan in case of a real fire?

Do you have all necessary equipment?

- ❏ Sound pressure level meter
- ❏ Volt-ohm meter
- ❏ Sensitivity tester
- ❏ Aerosol smoke (where permitted)
- ❏ Tools
- ❏ Copy of *NFPA 72, National Fire Alarm Code*

Are you limiting testing to specific areas to minimize disruption?

Do you have a list of devices/appliances that require testing?

Is there a releasing system present?

Are the service personnel aware of the hazards associated with the releasing system?

AFTER TESTING

Have the occupants been notified of the completion of testing?

Has the supervising station been notified of the completion of testing?

Has the fire department been notified of the completion of testing?

Does the installed system match the record drawings?

- ❏ If not, has any discrepancy been noted?
- ❏ If not, has a repair and re-test been conducted?

Has the owner been notified (verbally and in writing) of any defects or malfunctions?

Continued

Exhibit VI.1 Test Checklist *Continued*

Has the system been returned to normal operation?

Have any releasing systems been restored to normal?

Has an Inspection and Testing Form been completed?

- ❑ Copy given to Authority Having Jurisdiction?
- ❑ Copy given to owner?
- ❑ Copy retained for records?

Exhibit VI.2 Test Methods

Device	Method
1. Control Equipment	
a. Functions	At a minimum, control equipment shall be tested to verify correct receipt of alarm, supervisory, and trouble signals (inputs), operation of evacuation signals and auxiliary functions (outputs), circuit supervision including detection of open circuits and ground faults, and power supply supervision for detection of loss of ac power and disconnection of secondary batteries.
b. Fuses	The rating and supervision shall be verified.
c. Interfaced equipment	Integrity of single or multiple circuits providing interface between two or more control panels shall be verified. Interfaced equipment connections shall be tested by operating or simulating operation of the equipment being supervised. Signals required to be transmitted shall be verified at the control panel.
d. Lamps and LEDs	Lamps and LEDs shall be illuminated.
e. Primary (main) power supply	All secondary (standby) power shall be disconnected and tested under maximum load, including all alarm appliances requiring simultaneous operation. All secondary (standby) power shall be reconnected at end of test. For redundant power supplies, each shall be tested separately.

Continued

Exhibit VI.2 Test Methods *Continued*

Device	Method
2. Engine-Driven Generator	If an engine-driven generator dedicated to the fire alarm system is used as a required power source, operation of the generator shall be verified in accordance with NFPA 110, *Standard for Emergency and Standby Power Systems*, by the building owner.
3. Secondary (Standby) Power Supply	All primary (main) power supplies shall be disconnected and the occurrence of required trouble indication for loss of primary power shall be verified. The system's standby and alarm current demand shall be measured or verified and, using manufacturer's data, the ability of batteries to meet standby and alarm requirements shall be verified. General alarm systems shall be operated for a minimum of 5 minutes and emergency voice communications systems for a minimum of 15 minutes. Primary (main) power supply shall be reconnected at end of test.
4. Uninterrupted Power Supply (UPS)	If a UPS system dedicated to the fire alarm system is used as a required power source, operation of the UPS system shall be verified by the building owner in accordance with NFPA 111, *Standard on Stored Electrical Energy Emergency and Standby Power Systems*.
5. Batteries—General Tests	Prior to conducting any battery testing, the person conducting the test shall ensure that all system software stored in volatile memory is protected from loss.
a. Visual inspection	Batteries shall be inspected for corrosion or leakage. Tightness of connections shall be checked and ensured. If necessary, battery terminals or connections shall be cleaned and coated. Electrolyte level in lead acid batteries shall be visually inspected.

b. Battery replacement

Batteries shall be replaced in accordance with the recommendations of the alarm equipment manufacturer or when the recharged battery voltage or current falls below the manufacturer's recommendations.

c. Charger test

Operation of battery charger shall be checked in accordance with charger test for the specific type of battery.

d. Discharge test

With the battery charger disconnected, the batteries shall be load tested following the manufacturer's recommendations. The voltage level shall not fall below the levels specified.

Exception: An artificial load equal to the full fire alarm load connected to the battery shall be permitted to be used in conducting this test.

e. Load voltage test

With the battery charger disconnected, the terminal voltage shall be measured while supplying the maximum load required by its application.

The voltage level shall not fall below the levels specified for the specific type of battery. If the voltage falls below the level specified, corrective action shall be taken and the batteries shall be retested.

Exception: An artificial load equal to the full fire alarm load connected to the battery shall be permitted to be used in conducting this test.

6. Battery Tests (Specific Types)

a. Primary battery load voltage test

The maximum load for a No. 6 primary battery shall not be more than 2 amperes per cell. An individual (1.5-volt) cell shall be replaced when a load of 1 ohm reduces the voltage below 1 volt. A 6-volt assembly shall be replaced when a test load of 4 ohms reduces the voltage below 4 volts.

Continued

Exhibit VI.2 Test Methods *Continued*

Device	Method
b. Lead acid type	
1. Charger test	With the batteries fully charged and connected to the charger, the voltage across the batteries shall be measured with a voltmeter. The voltage shall be 2.30 volts per cell ±0.02 volts at 25°C (77°F) or as specified by the equipment manufacturer.
2. Load voltage test	Under load, the battery shall not fall below 2.05 volts per cell.
3. Specific gravity	The specific gravity of the liquid in the pilot cell or all of the cells shall be measured as required. The specific gravity shall be within the range specified by the manufacturer. Although the specified specific gravity varies from manufacturer to manufacturer, a range of 1.205–1.220 is typical for regular lead-acid batteries, while 1.240–1.260 is typical for high-performance batteries. A hydrometer that shows only a pass or fail condition of the battery and does not indicate the specific gravity shall not be used, because such a reading does not give a true indication of the battery condition.
c. Nickel cadmium type	
1. Charger test	With the batteries fully charged and connected to the charger, an ampere meter shall be placed in series with the battery under charge. The charging current shall be in accordance with the manufacturer's recommendations for the type of battery used. In the absence of specific information, 1/30 to 1/25 of the battery rating shall be used.
2. Load voltage test	Under load, the float voltage for the entire battery shall be 1.42 volts per cell, nominal. If possible, cells shall be measured individually.

d. Sealed lead acid type	
1. Charger test	With the batteries fully charged and connected to the charger, the voltage across the batteries shall be measured with a voltmeter. The voltage shall be 2.30 volts per cell ±0.02 volts at 25°C (77°F) or as specified by the equipment manufacturer.
2. Load voltage test	Under load, the battery shall perform in accordance with the battery manufacturer's specifications.
7. Public Reporting System Tests	In addition to the tests and inspection required above, the following requirements shall apply. Manual tests of the power supply for public reporting circuits shall be made and recorded at least once during each 24-hour period. Such tests shall include the following:
	a. Current strength of each circuit. Changes in current of any circuit exceeding 10 percent shall be investigated immediately.
	b. Voltage across terminals of each circuit inside of terminals of protective devices. Changes in voltage of any circuit exceeding 10 percent shall be investigated immediately.
	c. * Voltage between ground and circuits. If this test shows a reading in excess of 50 percent of that shown in the test specified in (b), the trouble shall be immediately located and cleared. Readings in excess of 25 percent shall be given early attention. These readings shall be taken with a calibrated voltmeter of not more than 100-ohms resistance per volt. Systems in which each circuit is supplied by an independent current source (Forms 3 and 4) require tests between ground and each side of each circuit. Common current source systems (Form 2) require voltage tests between ground and each terminal of each battery and other current source.

Continued

Exhibit VI.2 Test Methods *Continued*

Device	Method
	d. Ground current reading shall be permitted in lieu of (c). If this method of testing is used, all grounds showing a current reading in excess of 5 percent of the supplied line current shall be given immediate attention.
	e. Voltage across terminals of common battery, on switchboard side of fuses.
	f. Voltage between common battery terminals and ground. Abnormal ground readings shall be investigated immediately.
	Tests specified in (e) and (f) shall apply only to those systems using a common battery. If more than one common battery is used, each common battery shall be tested.
8. Transient Suppressors	Lightning protection equipment shall be inspected and maintained per the manufacturer's specifications.
	Additional inspections shall be required after any lightning strikes.
	Equipment located in moderate to severe areas outlined in NFPA 780, *Standard for the Installation of Lightning Protection Systems,* Appendix H, shall be inspected semiannually and after any lightning strikes.
9. Control Unit Trouble Signals	
a. Audible and visual	Operation of panel trouble signals shall be verified as well as ring-back feature for systems using a trouble-silencing switch that requires resetting.

b. Disconnect switches

If control unit has disconnect or isolating switches, performance of intended function of each switch shall be verified and receipt of trouble signal when a supervised function is disconnected shall also be verified.

c. Ground-fault monitoring circuit

If the system has a ground detection feature, the occurrence of ground-fault indication shall be verified whenever any installation conductor is grounded.

d. Transmission of signals to off-premises location

An initiating device shall be actuated and receipt of alarm signal at the off-premises location shall be verified.

A trouble condition shall be created and receipt of a trouble signal at the off-premises location shall be verified.

A supervisory device shall be actuated and receipt of a supervisory signal at the off-premises location shall be verified. If a transmission carrier is capable of operation under a single or multiple fault condition, an initiating device shall be actuated during such fault condition and receipt of a trouble signal at the off-premises location shall be verified, in addition to the alarm signal.

10. Remote Annunciators	The correct operation and identification of annunciators shall be verified. If provided, the correct operation of annunciator under a fault condition shall be verified.
11. Conductors/Metallic	
a. Stray voltage	All installation conductors shall be tested with a volt/ohmmeter to verify that there are no stray (unwanted) voltages between installation conductors or between installation conductors and ground. Unless a different threshold is specified in the system installed equipment manufacturer's specifications, the maximum allowable stray voltage shall not exceed 1 volt ac/dc.

Continued

Exhibit VI.2 Test Methods *Continued*

Device	Method
b. Ground faults	All installation conductors other than those intentionally and permanently grounded shall be tested for isolation from ground per the installed equipment manufacturer's specifications.
c. Short-circuit faults	All installation conductors other than those intentionally connected together shall be tested for conductor-to-conductor isolation per the installed equipment manufacturer's specifications. These same circuits also shall be tested conductor-to-ground.
d. Loop resistance	With each initiating and indicating circuit installation conductor pair short-circuited at the far end, the resistance of each circuit shall be measured and recorded. It shall be verified that the loop resistance does not exceed the installed equipment manufacturer's specified limits.
12. Conductors/Nonmetallic	
a. Circuit integrity	Each initiating device, notification appliance, and signaling line circuit shall be tested to confirm that the installation conductors are monitored for integrity in accordance with the requirements of Chapters 1 and 3.
b. Fiber optics	The fiber-optic transmission line shall be tested in accordance with the manufacturer's instructions by the use of an optical power meter or by an optical time domain reflectometer used to measure the relative power loss of the line. This relative figure for each fiber-optic line shall be recorded in the fire alarm control panel. If the power level drops 2 percent or more from the value recorded during the initial acceptance test, the transmission line, section thereof, or connectors shall be repaired or replaced by a qualified technician to bring the line back into compliance with the accepted transmission level per the manufacturer's recommendations.

c. Supervision

Introduction of a fault in any supervised circuit shall result in a trouble indication at the control unit. One connection shall be opened at not less than 10 percent of the initiating device, notification appliance, and signaling line circuit.

Each initiating device, notification appliance, and signaling line circuit shall be tested for correct indication at the control unit. All circuits shall perform as indicated in Table 3-5, Table 3-6, or Table 3-7.

13. Initiating Devices

a. Electromechanical releasing device

1. Nonrestorable-type link

Correct operation shall be verified by removal of the fusible link and operation of the associated device. Any moving parts shall be lubricated as necessary.

2. Restorable-type link*

Correct operation shall be verified by removal of the fusible link and operation of the associated device. Any moving parts shall be lubricated as necessary.

b. Fire extinguishing system(s) or suppression system(s) alarm switch

The switch shall be mechanically or electrically operated and receipt of signal by the control panel shall be verified.

c. Fire–gas and other detectors

Fire–gas detectors and other fire detectors shall be tested as prescribed by the manufacturer and as necessary for the application.

d. Heat detectors

Continued

Exhibit VI.2 Test Methods *Continued*

Device	Method
1. Fixed-temperature, rate-of-rise, rate-of-compensation, restorable line, spot type (excluding pneumatic tube type)	Heat test shall be performed with a heat source per the manufacturer's recommendations for response within 1 minute. A test method shall be used that is recommended by the manufacturer or other method shall be used that will not damage the nonrestorable fixed-temperature element of a combination rate-of-rise/fixed-temperature element.
2. Fixed-temperature, nonrestorable line type	Heat test shall not be performed. Functionality shall be tested mechanically and electrically. Loop resistance shall be measured and recorded. Changes from acceptance test shall be investigated.
3. Fixed-temperature, nonrestorable spot type	After 15 years from initial installation, all devices shall be replaced or two detectors per 100 shall be laboratory tested. The two detectors shall be replaced with new devices. If a failure occurs on any of the detectors removed, additional detectors shall be removed and tested to determine either a general problem involving faulty detectors or a localized problem involving one or two defective detectors. If detectors are tested instead of replaced, tests shall be repeated at intervals of 5 years.
4. Nonrestorable (general)	Heat tests shall not be performed. Functionality shall be tested mechanically and electrically.
5. Restorable line type, pneumatic tube only	Heat tests shall be performed (where test chambers are in circuit) or a test with pressure pump shall be conducted.

e. Fire alarm boxes

Manual fire alarm boxes shall be operated per the manufacturer's instructions. Key-operated presignal and general alarm manual fire alarm boxes shall both be tested.

f. Radiant energy fire detectors

Flame detectors and spark/ember detectors shall be tested in accordance with the manufacturer's instructions to determine that each detector is operative.

Flame detector and spark/ember detector sensitivity shall be determined using any of the following:

a. Calibrated test method

b. Manufacturer's calibrated sensitivity test instrument

c. Listed control unit arranged for the purpose

d. Other approved calibrated sensitivity test method that is directly proportional to the input signal from a fire, consistent with the detector listing or approval

If designed to be field adjustable, detectors found to be outside of the approved range of sensitivity shall be replaced or adjusted to bring them into the approved range.

Flame detector and spark/ember detector sensitivity shall not be determined using a light source that administers an unmeasured quantity of radiation at an undefined distance from the detector.

g. Smoke detectors

1. Systems detectors

The detectors shall be tested in place to ensure smoke entry into the sensing chamber and an alarm response. Testing with smoke or listed aerosol approved by the manufacturer shall be permitted as acceptable test methods. Other methods approved by the manufacturer that ensure smoke entry into the sensing chamber shall be permitted.

Continued

Exhibit VI.2 Test Methods *Continued*

Device	Method
	Any of the following tests shall be performed to ensure that each smoke detector is within its listed and marked sensitivity range:
	a. Calibrated test method
	b. Manufacturer's calibrated sensitivity test instrument
	c. Listed control equipment arranged for the purpose
	d. Smoke detector/control unit arrangement whereby the detector causes a signal at the control unit when its sensitivity is outside its listed sensitivity range
	e. Other calibrated sensitivity test method approved by the authority having jurisdiction
2. Single station detectors	The detectors shall be tested in place to ensure smoke entry into the sensing chamber and an alarm response. Testing with smoke or listed aerosol approved by the manufacturer shall be permitted as acceptable test methods. Other methods approved by the manufacturer that ensure smoke entry into the sensing chamber shall be permitted.
3. Air sampling	Per manufacturer's recommended test methods, detector alarm response shall be verified through the end sampling port on each pipe run; airflow through all other ports shall be verified as well.

4. Duct type
Air duct detectors shall be tested or inspected to ensure that the device will sample the air-stream. The test shall be made in accordance with the manufacturer's instructions.

5. Projected beam type
The detector shall be tested by introducing smoke, other aerosol, or an optical filter into the beam path

6. Smoke detector with built-in thermal element
Both portions of the detector shall be operated independently as described for the respective devices.

7. Smoke detectors with control output functions
It shall be verified that the control capability shall remain operable even if all of the initiating devices connected to the same initiating device circuit or signaling line circuit are in an alarm state.

h. Initiating devices, supervisory

1. Control valve switch
Valve shall be operated and signal receipt shall be verified to be within the first two revolutions of the hand wheel or within one-fifth of the travel distance, or per the manufacturer's specifications.

2. High- or low-air pressure switch
Switch shall be operated. Receipt of signal obtained where the required pressure is increased or decreased a maximum 10 psi (70 kPa) from the required pressure level shall be verified.

Continued

METHODS AND FREQUENCIES

Exhibit VI.2 Test Methods *Continued*

Device	Method
3. Room temperature switch	Switch shall be operated. Receipt of signal to indicate the decrease in room temperature to 40°F (4.4°C) and its restoration to above 40°F (4.4°C) shall be verified.
4. Water level switch	Switch shall be operated. Receipt of signal indicating the water level raised or lowered 3 in. (76.2 mm) from the required level within a pressure tank, or 12 in. (305 mm) from the required level of a nonpressure tank, shall be verified, as shall its restoral to required level.
5. Water temperature switch	Switch shall be operated. Receipt of signal to indicate the decrease in water temperature to 40°F (4.4°C) and its restoration to above 40°F (4.4°C) shall be verified.
i. Mechanical, electrosonic, or pressure-type waterflow device	Water shall be flowed through an inspector's test connection indicating the flow of water equal to that from a single sprinkler of the smallest orifice size installed in the system for wet-pipe systems, or an alarm test bypass connection for dry-pipe, pre-action, or deluge systems in accordance with NFPA 25, *Standard for the Inspection, Testing, and Maintenance of Water-Based Fire Protection Systems.*
14. Alarm Notification Appliances	
a. Audible	Sound pressure level shall be measured with sound level meter meeting ANSI S1.4a, *Specifications for Sound Level Meters,* Type 2 requirements. Levels throughout protected area shall be measured and recorded.

b. Audible textural notification appliances (speakers and other appliances to convey voice messages)

Sound pressure level shall be measured with sound level meter meeting ANSI S1.4a, *Specifications for Sound Level Meters*, Type 2 requirements. Levels throughout protected area shall be measured and recorded.

c. Visible

Audible information shall be verified to be distinguishable and understandable.

Test shall be performed in accordance with the manufacturer's instructions. Device locations shall be verified to be per approved layout and it shall be confirmed that no floor plan changes affect the approved layout.

15. Special Hazard Equipment

a. Abort switch (IRI type)

Abort switch shall be operated. Correct sequence and operation shall be verified.

b. Abort switch (recycle type)

Abort switch shall be operated. Development of correct matrix with each sensor operated shall be verified.

c. Abort switch (special type)

Abort switch shall be operated. Correct sequence and operation in accordance with authority having jurisdiction shall be verified. Sequencing on as-built drawings or in owner's manual shall be observed.

d. Cross zone detection circuit

One sensor or detector on each zone shall be operated. Occurrence of correct sequence with operation of first zone and then with operation of second zone shall be verified.

Continued

Exhibit VI.2 Test Methods *Continued*

Device	Method
e. Matrix-type circuit	All sensors in system shall be operated. Development of correct matrix with each sensor operated shall be verified.
f. Release solenoid circuit	Solenoid shall be used with equal current requirements. Operation of solenoid shall be verified.
g. Squib release circuit	AGI flashbulb or other test light approved by the manufacturer shall be used. Operation of flashbulb or light shall be verified.
h. Verified, sequential, or counting zone circuit	Required sensors at a minimum of four locations in circuit shall be operated. Correct sequence with both the first and second detector in alarm shall be verified.
i. All above devices or circuits or combinations thereof	Supervision of circuits shall be verified by creating an open circuit.

16. Supervising Station Fire Alarm Systems—Transmission Equipment

a. All equipment	Test shall be performed on all system functions and features in accordance with the equipment manufacturer's instructions for correct operation in conformance with the applicable sections of Chapter 5.
	Initiating device shall be actuated. Receipt of the correct initiating device signal at the supervising station within 90 seconds shall be verified. Upon completion of the test, the system shall be restored to its functional operating condition.

b. Digital alarm communicator transmitter (DACT)

If test jacks are used, the first and last tests shall be made without the use of the test jack.

Connection of the DACT to two separate means of transmission shall be ensured. *Exception: DACTs that are connected to a telephone line (number) that is also supervised for adverse conditions by a derived local channel.*

DACT shall be tested for line seizure capability by initiating a signal while using the primary line for a telephone call. Receipt of the correct signal at the supervising station shall be verified. Completion of the transmission attempt within 90 seconds from going off-hook to on-hook shall be verified.

The primary line from the DACT shall be disconnected. Indication of the DACT trouble signal at the premises shall be verified as well as transmission to the supervising station within 4 minutes of detection of the fault.

The secondary means of transmission from the DACT shall be disconnected. Indication of the DACT trouble signal at the premises shall be verified as well as transmission to the supervising station within 4 minutes of detection of the fault.

The DACT shall be caused to transmit a signal to the DACR while a fault in the primary telephone number is simulated. Utilization of the secondary telephone number by the DACT to complete the transmission to the DACR shall be verified.

c. Digital alarm radio transmitter (DART)

The primary telephone line shall be disconnected. Transmission of a trouble signal to the supervising station by the DART within 4 minutes shall be verified.

Continued

Exhibit VI.2 Test Methods *Continued*

Device	Method
d. McCulloh transmitter	Initiating device shall be actuated. Production of not less than three complete rounds of not less than three signal impulses each by the McCulloh transmitter shall be verified.
	If end-to-end metallic continuity is present and with a balanced circuit, each of the following four transmission channel fault conditions shall be caused in turn, and receipt of correct signals at the supervising station shall be verified:
	a. Open
	b. Ground
	c. Wire-to-wire short
	d. Open and ground
	If end-to-end metallic continuity is not present and with a properly balanced circuit, each of the following three transmission channel fault conditions shall be caused in turn, and receipt of correct signals at the supervising station shall be verified
	a. Open
	b. Ground
	c. Wire-to-wire short

e. Radio alarm transmitter (RAT)	A fault between elements of the transmitting equipment shall be caused. Indication of the fault at the protected premises shall be verified or it shall be verified that a trouble signal is transmitted to the supervising station.

17. Supervising Station Fire Alarm Systems—Receiving Equipment

a. All Equipment	Tests shall be performed on all system functions and features in accordance with the equipment manufacturer's instructions for correct operation in conformance with the applicable sections of Chapter 5.
	Initiating device shall be actuated. Receipt of the correct initiating device signal at the supervising station within 90 seconds shall be verified. Upon completion of the test, the system shall be restored to its functional operating condition.
	If test jacks are used, the first and last tests shall be made without the use of the test jack.
b. Digital alarm communicator receiver (DACR)	Each telephone line (number) shall be disconnected in turn from the DACR and audible and visual annunciation of a trouble signal in the supervising station shall be verified.
	A signal shall be caused to be transmitted on each individual incoming DACR line at least once every 24 hours. Receipt of these signals shall be verified.
c. Digital alarm radio receiver (DARR)	The following conditions of all DARRs on all subsidiary and repeater station receiving equipment shall be caused. Receipt at the supervising station of correct signals for each of the following conditions shall be verified:
	a. AC power failure of the radio equipment
	b. Receiver malfunction

Continued

Exhibit VI.2 Test Methods *Continued*

Device	Method
	c. Antenna and interconnecting cable failure
	d. Indication of automatic switchover of the DARR
	e. Data transmission line failure between the DARR and the supervising or subsidiary station
	The current on each circuit at each supervising and subsidiary station under the following conditions shall be tested and recorded:
	a. During functional operation
	b. On each side of the circuit with the receiving equipment conditioned for an open circuit
	A single break or ground condition shall be caused on each transmission channel. If such a fault prevents the functioning of the circuit, receipt of a trouble signal shall be verified.
	Each of the following conditions at each of the supervising or subsidiary stations and all re-peater station radio transmitting and receiving equipment shall be caused; receipt of correct signals at the supervising station shall be verified:
d. McCulloh systems	a. RF transmitter in use (radiating)
	b. AC power failure supplying the radio equipment
	c. RF receiver malfunction
	d. Indication of automatic switchover

e. Radio alarm supervising station receiver (RASSR) and radio alarm repeater station receiver (RARSR)	Each of the following conditions at each of the supervising or subsidiary stations and all repeater station radio transmitting and receiving equipment shall be caused; receipt of correct signals at the supervising station shall be verified: a. AC power failure supplying the radio equipment b. RF receiver malfunction c. Indication of automatic switchover, if applicable
f. Private microwave radio systems	Each of the following conditions at each of the supervising or subsidiary stations and all repeater station radio transmitting and receiving equipment shall be caused; receipt of correct signals at the supervising station shall be verified: a. RF transmitter in use (radiating) b. AC power failure supplying the radio equipment c. RF receiver malfunction d. Indication of automatic switchover
18. Emergency Communications Equipment	
a. Amplifier/tone generators	Correct switching and operation of backup equipment shall be verified.
b. Call-in signal silence	Function shall be operated and receipt of correct visual and audible signals at control panel shall be verified.

Continued

Exhibit VI.2 Test Methods *Continued*

Device	Method
c. Off-hook indicator (ring down)	Phone set shall be installed or phone shall be removed from hook and receipt of signal at control panel shall be verified.
d. Phone jacks	Phone jack shall be visually inspected and communications path through jack shall be initiated.
e. Phone set	Each phone set shall be activated and correct operation shall be verified.
f. System performance	System shall be operated with a minimum of any five handsets simultaneously. Voice quality and clarity shall be verified.
19. Interface Equipment	Interface equipment connections shall be tested by operating or simulating the equipment being supervised. Signals required to be transmitted shall be verified at the control panel. Test frequency for interface equipment shall be the same as the frequency required by the applicable NFPA standard(s) for the equipment being supervised.
20. Guard's Tour Equipment	The device shall be tested in accordance with the manufacturer's specifications.
21. Special Procedures	
a. Alarm verification	Time delay and alarm response for smoke detector circuits identified as having alarm verification shall be verified.

b. Multiplex systems

> Communications between sending and receiving units under both primary and secondary power shall be verified.
>
> Communications between sending and receiving units under open circuit and short circuit trouble conditions shall be verified.
>
> Communications between sending and receiving units in all directions where multiple communications pathways are provided shall be verified.
>
> If redundant central control equipment is provided, switchover and all required functions and operations of secondary control equipment shall be verified.
>
> All system functions and features shall be verified in accordance with manufacturer's instructions.

22. Low-Power Radio (Wireless Systems)

> The following procedures describe additional acceptance and reacceptance test methods to verify wireless protection system operation:
>
> a. The manufacturer's manual and the as-built drawings provided by the system supplier shall be used to verify correct operation after the initial testing phase has been performed by the supplier or by the supplier's designated representative.
>
> *Continued*

Exhibit VI.2 Test Methods *Continued*

Device	Method
	b. Starting from the functional operating condition, the system shall be initialized in accordance with the manufacturer's manual. A test shall be conducted to verify the alternative path, or paths, by turning off or disconnecting the primary wireless repeater. The alternative communications path shall exist between the wireless control panel and peripheral devices used to establish initiation, indicating, control, and annunciation. The system shall be tested for both alarm and trouble conditions.
	c. Batteries for all components in the system shall be checked monthly. If the control panel checks all batteries and all components daily, the system shall not require monthly testing of the batteries.

Source: NFPA 72®, National Fire Alarm Code®, NFPA, Quincy, MA, 1999 edition, Table 7-2.2.

Editor's Note: An asterisk (*) following a paragraph number or component indicates nonmandatory material appeared in Appendix A of the original document. For the reader's convenience, this material has been merged into the table. Cross-references refer to NFPA 72.

Exhibit VI.3 Testing Frequencies

Component	Initial/ Reacceptance	Monthly	Quarterly	Semiannually	Annually
1. Control Equipment—Building Systems Connected to Supervising Station					
a. Functions	X	—	—	—	X
b. Fuses	X	—	—	—	X
c. Interfaced equipment	X	—	—	—	X
d. Lamps and LEDs	X	—	—	—	X
e. Primary (main) power supply	X	—	—	—	X
f. Transponders	X	—	—	—	X
2. Control Equipment—Building Systems Not Connected to a Supervising Station	—			—	—
a. Functions	X	—	X	—	—
b. Fuses	X	—	X	—	—
c. Interfaced equipment	X	—	X	—	—
d. Lamps and LEDs	X	—	X	—	—
e. Primary (main) power supply	X	—	X	—	—
f. Transponders	X	—	X	—	—

Continued

Exhibit VI.3 Testing Frequencies *Continued*

Component	Initial/Reacceptance	Monthly	Quarterly	Semiannually	Annually
3. Engine-Driven Generator—Central Station Facilities and Fire Alarm Systems	X	X	—	—	—
4. Engine-Driven Generator—Public Fire Alarm Reporting Systems	X	—	—	—	—
5. Batteries—Central Station Facilities					
a. Lead acid type					
1. Charger test (Replace battery as needed.)	—	—	—	—	X
2. Discharge test (30 min)	X	—	—	—	X
3. Load voltage test	X	—	—	X	—
4. Specific gravity	X	X	—	—	—
b. Nickel cadmium type					
1. Charger test (Replace battery as needed.)	—	—	—	—	X
2. Discharge test (30 min)	X	—	—	—	X
3. Load voltage test	X	X	—	—	—
c. Sealed lead acid type					
1. Charger test (Replace battery as needed.)	—	—	X	—	—

Continued

	1	2	3	4	5
2. Discharge test (30 min)	×	×			
3. Load voltage test	×	×			
6. Batteries—Fire Alarm Systems					
a. Lead acid type					
1. Charger test (Replace battery as needed.)	×			×	×
2. Discharge test (30 min)	×			×	
3. Load voltage test	×			×	
4. Specific gravity					
b. Nickel cadmium type					
1. Charger test (Replace battery as needed.)	×			×	×
2. Discharge test (30 min)	×				×
3. Load voltage test	×	×			
c. Primary type (dry cell)					
1. Load voltage test	×				
d. Sealed lead acid type					
1. Charger test (Replace battery every 4 yr.)	×				×
2. Discharge test (30 min)	×				×
3. Load voltage test	×			×	

Exhibit VI.3 Testing Frequencies *Continued*

Component	Initial/Reacceptance	Monthly	Quarterly	Semiannually	Annually
7. Batteries—Public Fire Alarm Reporting Systems Voltage Tests in Accordance with Table 7-2.2, item 7(a)–(f)	X (daily)	—	—	—	—
a. Lead acid type					
1. Charger test (Replace battery as needed.)	×				×
2. Discharge test (2 hr)	×		×		
3. Load voltage test	×		×		
4. Specific gravity	×			×	
b. Nickel cadmium type					
1. Charger test (Replace battery as needed.)	×				×
2. Discharge test (2 hr)	×		×		
3. Load voltage test	×				×
c. Sealed lead acid type					
1. Charger test (Replace battery as needed.)	×				×
2. Discharge test (2 hr)	×		×		
3. Load voltage test	×				×
8. Fiber-Optic Cable Power	×				×

9. Control Unit Trouble Signals	X	—	—	X
10. Conductors—Metallic	X	—	—	—
11. Conductors—Nonmetallic	X	—	—	—
12. Emergency Voice/Alarm Communications Equipment	X	—	—	X
13. Retransmission Equipment (The requirements of 7-3.4 shall apply.)	X	—	—	
14. Remote Annunciators	X	—	—	X
15. Initiating Devices	—	—	—	—
a. Duct detectors	X	—	—	X
b. Electromechanical releasing device	X	—	—	X
c. Fire-extinguishing system(s) or suppression system(s) switches	X	—	—	X
d. Fire–gas and other detectors	X	—	—	X
e. Heat detectors (The requirements of 7-3.2.3 shall apply.)	X	—	—	X
f. Fire alarm boxes	X	—	—	X
g. Radiant energy fire detectors	X	X	—	—
h. All smoke detectors—functional	X	—	—	X

Continued

Exhibit VI.3 Testing Frequencies *Continued*

Component	Initial/ Reacceptance	Monthly	Quarterly	Semiannually	Annually
i. Smoke detectors—sensitivity (The requirements of 7-3.2.1 shall apply.)	—	—	—	—	—
j. Supervisory signal devices (except valve tamper switches)	X	—	X	—	—
k. Waterflow devices	X	—	—	X	—
l. Valve tamper switches	X	—	—	X	—
16. Guard's Tour Equipment	X	—	—	—	X
17. Interface Equipment	X	—	—	—	X
18. Special Hazard Equipment	X	—	—	—	X
19. Alarm Notification Appliances	—	—	—	—	—
a. Audible devices	X	—	—	—	X
b. Audible textual notification appliances	X	—	—	—	X
c. Visible devices	X	—	—	—	X
20. Off-Premises Transmission Equipment	X	—	X	—	—

Item					
21. Supervising Station Fire Alarm Systems—Transmitters					
a. DACT	×				×
b. DART	×				×
c. McCulloh	×				×
d. RAT	×				×
22. Special Procedures	×				×
23. Supervising Station Fire Alarm Systems—Receivers					
a. DACR				×	×
b. DARR				×	×
c. McCulloh systems				×	×
d. Two-way RF multiplex				×	×
e. RASSR				×	×
f. RARSR				×	×
g. Private microwave				×	×

Source: NFPA 72®, National Fire Alarm Code®, NFPA, Quincy, MA, 1999 edition, Table 7-3.2.

Exhibit VI.4 Visual Inspection Frequencies

Component	Initial/ Reacceptance	Monthly	Quarterly	Semiannually	Annually
1. Control Equipment: Fire Alarm Systems Monitored for Alarm, Supervisory, Trouble Signals					
a. Fuses	X	—	—	—	X
b. Interfaced equipment	X	—	—	—	X
c. Lamps and LEDs	X	—	—	—	X
d. Primary (main) power supply	X	—	—	—	X
2. Control Equipment: Fire Alarm Systems Unmonitored for Alarm, Supervisory, Trouble Signals					
a. Fuses	X (weekly)	—	—	—	—
b. Interfaced equipment	X (weekly)	—	—	—	—
c. Lamps and LEDs	X (weekly)	—	—	—	—
d. Primary (main) power supply	X (weekly)	—	—	—	—
3. Batteries					
a. Lead acid	X	X	—	—	—
b. Nickel cadmium	X	—	—	X	—
c. Primary (dry cell)	X	X	—	—	—
d. Sealed lead acid	X	—	—	X	—
4. Transient Suppressors	X	—	—	X	—

Component	Col 1	Col 2	Col 3	Col 4	Col 5
5. Control Unit Trouble Signals	X (weekly)	—	—	X	—
6. Fiber-Optic Cable Connections	X	—	—	—	×
7. Emergency Voice/Alarm Communications Equipment	X	—	—	X	—
8. Remote Annunciators	X	—	—	X	—
9. Initiating Devices					
a. Air sampling	X	—	—	X	—
b. Duct detectors	X	—	—	X	—
c. Electromechanical releasing devices	X	—	—	X	—
d. Fire-extinguishing system(s) or suppression system(s) switches	X	—	—	X	—
e. Fire alarm boxes	X	—	—	X	—
f. Heat detectors	X	—	—	X	—
g. Radiant energy fire detectors	X	—	×	—	—
h. Smoke detectors	X	—	—	X	—
i. Supervisory signal devices	X	—	×	—	—
j. Waterflow devices	X	—	×	—	—
10. Guard's Tour Equipment	X	—	—	X	—
11. Interface Equipment	X	—	—	X	—

Continued

Exhibit VI.4 Visual Inspection Frequencies *Continued*

Component	Initial/ Reacceptance	Monthly	Quarterly	Semiannually	Annually
12. Alarm Notification Appliances—Supervised	X	—	—	X	—
13. Supervising Station Fire Alarm Systems—Transmitters					
a. DACT	X	—	—	X	—
b. DART	X	—	—	X	—
c. McCulloh	X	—	—	X	—
d. RAT	X	—	—	X	—
14. Special Procedures	X	—	—	X	—
15. Supervising Station Fire Alarm Systems—Receivers					
a. DACR[a]	X	X	—	—	—
b. DARR[a]	X	X	—	—	—
c. McCulloh systems[a]	X	—	—	X	—
d. Two-way RF multiplex[a]	X	—	—	X	—
e. RASSR[a]	X	—	—	X	—
f. RARS[a]	X	—	—	X	—
g. Private microwave[a]	X	—	—	X	—

[a]Reports of automatic signal receipt shall be verified daily.

Source: NFPA 72®, National Fire Alarm Code®, NFPA, Quincy, MA, 1999 edition, Table 7-3.1.

BATTERY VOLTAGE

Exhibit VI.5 Voltage for Nickel Cadmium Batteries

Float voltage	1.42 V/cell + 0.01 V
High rate voltage	1.58 V/cell + 0.07 V − 0.00 V

Note: High and low gravity voltages are (+) 0.07 V and (−) 0.03 V, respectively.

Source: NFPA 72®, National Fire Alarm Code®, NFPA, Quincy, MA, 1999 edition, Table A-7-2.2(c)(1).

Exhibit VI.6 Voltage for Lead Acid Batteries

Float Voltage	High-Gravity Battery (lead calcium)	Low-Gravity Battery (lead antimony)
Maximum	2.25 V/cell	2.17 V/cell
Minimum	2.20 V/cell	2.13 V/cell
High rate voltage	—	2.33 V/cell

Source: NFPA 72®, National Fire Alarm Code®, NFPA, Quincy, MA, 1999 edition, Table A-7-2.2(c)(2).

SECTION VII

WIRING

This section provides essential data for the designer and installer relevant to wiring methods, conductors, cable substitutions, and conduit and raceway systems. This section contains voltage drop and loop resistance information, conductor and cable data, conduit and box fill requirements, and specifications for power-limited, Class 2, and Class 3 power supplies.

For explanation, correct and incorrect wiring methods are illustrated and compared. A primer on commonly used electrical relationships including Ohm's law and Kirchoff's laws is included. Basic technical criteria, relationships, and formulas are presented for inductance, capacitance, and impedance. A resistor color code table is included.

This section includes numerous tables pertaining to dimensions and percent area of the most commonly used conduit and raceway systems.

Exhibit VII.1a Wiring Method
for Pigtail Connections—Correct

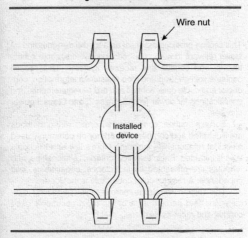

Wire nut

Installed device

Source: Richard W. Bukowski and Robert J. O'Laughlin, *Fire Alarm Signaling Systems,* 2nd ed., NFPA, Quincy, MA; Society of Fire Protection Engineers, Boston, 1994, Fig. 13-7.

**Exhibit VII.1b Wiring Method
for Pigtail Connections—Incorrect**

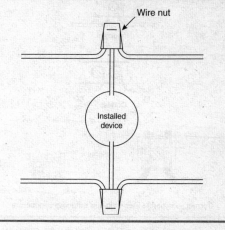

Source: Richard W. Bukowski and Robert J. O'Laughlin, *Fire Alarm Signaling Systems,* 2nd ed., NFPA, Quincy, MA; Society of Fire Protection Engineers, Boston, 1994, Fig. 13-6.

Exhibit VII.2 Wiring Methods for Terminations

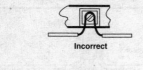

Incorrect

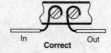

In   Out

Correct

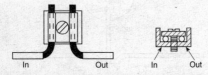

In   Out   In   Out

Correct — separate incoming and outgoing conductors

Source: NFPA 72®, National Fire Alarm Code®, NFPA, Quincy, MA, 1999 edition, Fig. A-2-1.3.4(a).

Exhibit VII.3 Wiring Methods

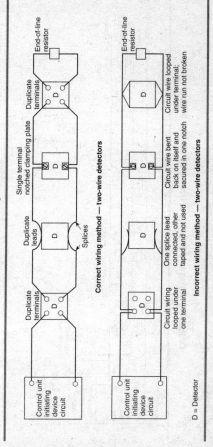

Correct wiring method — two-wire detectors

Incorrect wiring method — two-wire detectors

D = Detector

Source: NFPA 72®, National Fire Alarm Code®, NFPA, Quincy, MA, 1999 edition, Fig. A-2-1.3.4(a).

Exhibit VII.4 Wiring Arrangements for Three- and Four-Wire Detectors

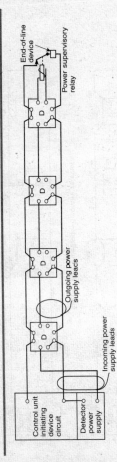

End-of-line device

Power supervisory relay

Outgoing power supply leads

Incoming power supply leads

Control unit initiating device circuit

Detector power supply

D = Detector

Illustrates four-wire smoke detector employing a three-wire connecting arrangement. One side of power supply is connected to one side of initiating device circuit. Wire run broken at each connection to smoke detector to provide supervision.

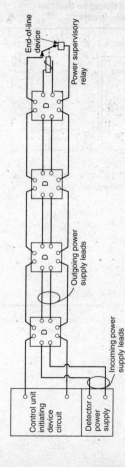

End-of-line device

Power supervisory relay

Outgoing power supply leads

Incoming power supply leads

Control unit initiating device circuit

Detector power supply

D = Detector

Illustrates four-wire smoke detector employing a four-wire connecting arrangement. Incoming and outgoing leads or terminals for both initiating device and power supply connections. Wire run broken at each connection to provide supervision.

Source: NFPA 72®, *National Fire Alarm Code*®, NFPA, Quincy, MA, 1999 edition, Fig. A-2-1.3.4(b).

Exhibit VII.5a Wiring Method for Multiriser
IDC Circuit—Correct

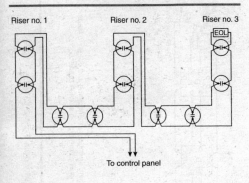

To control panel

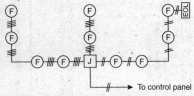

To control panel

Source: Richard W. Bukowski and Robert J. O'Laughlin, *Fire Alarm Signaling Systems,* 2nd ed., NFPA, Quincy, MA; Society of Fire Protection Engineers, Boston, 1994, Fig. 13-9.

Exhibit VII.5b Wiring Method for Multiriser IDC Circuit—Incorrect

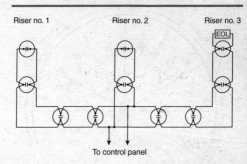

Riser no. 1   Riser no. 2   Riser no. 3

To control panel

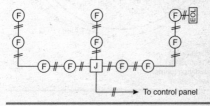

To control panel

Source: Richard W. Bukowski and Robert J. O'Laughlin, *Fire Alarm Signaling Systems,* 2nd ed., NFPA, Quincy, MA; Society of Fire Protection Engineers, Boston, 1994, Fig.13-8.

Exhibit VII.6a Basic Electrical Ohm's Law Relationships

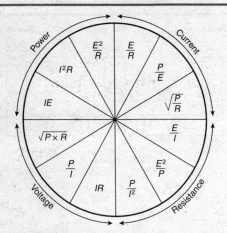

E = Voltage (volts)
I = Current (amps)
R = Resistance (ohms)
P = Power (watts)

ELECTRICAL RELATIONSHIPS

Exhibit VII.6b Basic Electrical Ohm's Law Relationships

$$I = \sqrt{\frac{P}{R}} \qquad \text{Amps} = \sqrt{\frac{\text{Watts}}{\text{Ohms}}}$$

$$I = \frac{P}{E} \qquad \text{Amps} = \frac{\text{Watts}}{\text{Volts}}$$

$$I = \frac{E}{R} \qquad \text{Amps} = \frac{\text{Volts}}{\text{Ohms}}$$

- -

$$P = \frac{E^2}{R} \qquad \text{Watts} = \frac{\text{Volts}^2}{\text{Ohms}}$$

$$P = E \times I \qquad \text{Watts} = \text{Volts} \times \text{Amps}$$

$$P = I^2 \times R \qquad \text{Watts} = \text{Amps}^2 \times \text{Ohms}$$

- -

$$E = \sqrt{P \times R} \qquad \text{Volts} = \sqrt{\text{Watts} \times \text{Ohms}}$$

$$E = I \times R \qquad \text{Volts} = \text{Amps} \times \text{Ohms}$$

$$E = \frac{P}{I} \qquad \text{Volts} = \frac{\text{Watts}}{\text{Amps}}$$

- -

$$R = \frac{E^2}{P} \qquad \text{Ohms} = \frac{\text{Volts}^2}{\text{Watts}}$$

$$R = \frac{P}{I^2} \qquad \text{Ohms} = \frac{\text{Watts}}{\text{Amps}^2}$$

$$R = \frac{E}{I} \qquad \text{Ohms} = \frac{\text{Volts}}{\text{Amps}}$$

- -

Editor's Note: Ohm's law is a mathematical relationship between current, voltage, resistance, and power. The current flow in a circuit is directly proportional to the voltage and inversely proportional to the resistance. The current is the flow of charge; the potential difference (voltage) between two points is the cause of the current flow; and the opposition to the current flow is the resistance. Power is an indicator of work accomplished over time. The electrical measurement for power is the watt, which is equal to 1 joule/second. The power delivered by an energy source to an electrical device is equal to the product of the source voltage and the current through that source.

ELECTRICAL RELATIONSHIPS

Exhibit VII.7 Kirchoff's Current Law

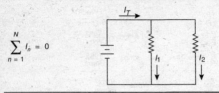

$$\sum_{n=1}^{N} I_n = 0$$

Editor's Note: The sum of the currents arriving at any point in a circuit must equal the sum of the currents leaving that point.

Exhibit VII.8 Kirchoff's Voltage Law

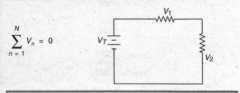

$$\sum_{n=1}^{N} V_n = 0$$

Editor's Note: The total voltage applied to any closed circuit path is always equal to the sum of the voltage drops in that path.

OR

The algebraic sum of all the voltages encountered in any closed loop equals zero.

ELECTRICAL RELATIONSHIPS

Exhibit VII.9 Inductance

Inductance (L) is the production of magnetization of electrification in a body by the proximity of a magnetic field or electric charge, or of the electric current in a conductor by the variation of the magnetic field in its vicinity, expressed in henrys.

In a series circuit:

$$L_T = L_1 + L_2 + L_3 + L_4$$

In a parallel circuit:

$$\frac{1}{L_T} = \frac{1}{L_1} + \frac{1}{L_2} + \frac{1}{L_3} + \frac{1}{L_4}$$

Exhibit VII.10 Capacitance

$$C = \frac{Q}{E} \qquad \text{Capacitance} = \frac{\text{Coulombs}}{\text{Volts}}$$

Capacitance (C) is the property of a circuit or body that permits it to store an electrical charge equal to the accumulated charge divided by the voltage, expressed in farads.

In a series circuit:

$$\frac{1}{C_T} = \frac{1}{C_1} + \frac{1}{C_2} + \frac{1}{C_3} + \frac{1}{C_4}$$

In a parallel circuit:

$$C_T = C_1 + C_2 + C_3 + C_4$$

A farad is the unit of capacitance of a condenser that retains one coulomb of charge with one volt difference of potential.

$$1 \text{ farad} = 1,000,000 \text{ microfarads}$$

ELECTRICAL RELATIONSHIPS

Exhibit VII.11 Impedance

Impedance (Z) is the total opposition to an alternating current presented by a circuit. Impedance is expressed in ohms.

$$Z = \frac{E}{I} \qquad \text{Impedance} = \frac{\text{Volts}}{\text{Amperes}}$$

Exhibit VII.12 Resistor Color Code Table

Color	First Band	Second Band	Third Band (Multiplier)	Fourth Band (Tolerance, %)
Black	0	0	1	—
Brown	1	1	10	—
Red	2	2	100	—
Orange	3	3	1,000	—
Yellow	4	4	10,000	—
Green	5	5	100,000	—
Blue	6	6	1,000,000	—
Violet	7	7	10,000,000	—
Gray	8	8	100,000,000	—
White	9	9	1,000,000,000	—
Gold	—	—	0.1	5
Silver	—	—	0.01	10
No color	—	—	—	20

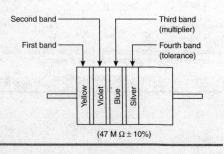

Second band — Third band (multiplier)

First band — Fourth band (tolerance)

Yellow | Violet | Blue | Silver

(47 M Ω ± 10%)

Exhibit VII.13 Conductor Properties

Size (AWG or kcmil)	Area		Stranding			Conductors				Direct-Current Resistance at 75°C (167°F)					
				Diameter		Overall				Copper				Aluminum	
						Diameter		Area		Uncoated		Coated			
	mm²	Circular mils	Qty	mm	in.	mm	in.	mm²	in.²	ohm/km	ohm/kFT	ohm/km	ohm/kFT	ohm/km	ohm/kFT
18	0.823	1620	1	—	—	1.02	0.040	0.823	0.001	25.5	7.77	26.5	8.08	42.0	12.8
18	0.823	1620	7	0.39	0.015	1.16	0.046	1.06	0.002	26.1	7.95	27.7	8.45	42.8	13.1
16	1.31	2580	1	—	—	1.29	0.051	1.31	0.002	16.0	4.89	16.7	5.08	26.4	8.05
16	1.31	2580	7	0.49	0.019	1.46	0.058	1.68	0.003	16.4	4.99	17.3	5.29	26.9	8.21
14	2.08	4110	1	—	—	1.63	0.064	2.08	0.003	10.1	3.07	10.4	3.19	16.6	5.06
14	2.08	4110	7	0.62	0.024	1.85	0.073	2.68	0.004	10.3	3.14	10.7	3.26	16.9	5.17
12	3.31	6530	1	—	—	2.05	0.081	3.31	0.005	6.34	1.93	6.57	2.01	10.45	3.18
12	3.31	6530	7	0.78	0.030	2.32	0.092	4.25	0.006	6.50	1.98	6.73	2.05	10.69	3.25
10	5.261	10380	1	—	—	2.588	0.102	5.26	0.008	3.984	1.21	4.148	1.26	6.561	2.00
10	5.261	10380	7	0.98	0.038	2.95	0.116	6.76	0.011	4.070	1.24	4.226	1.29	6.679	2.04
8	8.367	16510	1	—	—	3.264	0.128	8.37	0.013	2.506	0.764	2.579	0.786	4.125	1.26
8	8.367	16510	7	1.23	0.049	3.71	0.146	10.76	0.017	2.551	0.778	2.653	0.809	4.204	1.28

Continued

Exhibit VII.13 Conductor Properties *Continued*

Size (AWG or kcmil)	Area		Conductors							Direct-Current Resistance at 75°C (167°F)					
			Stranding			Overall				Copper				Aluminum	
				Diameter		Diameter		Area		Uncoated		Coated			
	mm²	Circular mils	Qty	mm	in.	mm	in.	mm²	in.²	ohm/km	ohm/kFT	ohm/km	ohm/kFT	ohm/km	ohm/kFT
6	13.30	26240	7	1.56	0.061	4.67	0.184	17.09	0.027	1.608	0.491	1.671	0.510	2.652	0.808
4	21.15	41740	7	1.96	0.077	5.89	0.232	27.19	0.042	1.010	0.308	1.053	0.321	1.666	0.508
3	26.67	52620	7	2.20	0.087	6.60	0.260	34.28	0.053	0.802	0.245	0.833	0.254	1.320	0.403
2	33.62	66360	7	2.47	0.097	7.42	0.292	43.23	0.067	0.634	0.194	0.661	0.201	1.045	0.319
1	42.41	83690	19	1.69	0.066	8.43	0.332	55.80	0.087	0.505	0.154	0.524	0.160	0.829	0.253
1/0	53.49	105600	19	1.89	0.074	9.45	0.372	70.41	0.109	0.399	0.122	0.415	0.127	0.660	0.201
2/0	67.43	133100	19	2.13	0.084	10.62	0.418	88.74	0.137	0.3170	0.0967	0.329	0.101	0.523	0.159
3/0	85.01	167800	19	2.39	0.094	11.94	0.470	111.9	0.173	0.2512	0.0766	0.2610	0.0797	0.413	0.126
4/0	107.2	211600	19	2.68	0.106	13.41	0.528	141.1	0.219	0.1996	0.0608	0.2050	0.0626	0.328	0.100
250	—	—	37	2.09	0.082	14.61	0.575	168	0.260	0.1687	0.0515	0.1753	0.0535	0.2778	0.0847
300	—	—	37	2.29	0.090	16.00	0.630	201	0.312	0.1409	0.0429	0.1463	0.0446	0.2318	0.0707
350	—	—	37	2.47	0.097	17.30	0.681	235	0.364	0.1205	0.0367	0.1252	0.0382	0.1984	0.0605
400	—	—	37	2.64	0.104	18.49	0.728	268	0.416	0.1053	0.0321	0.1084	0.0331	0.1737	0.0529
500	—	—	37	2.95	0.116	20.65	0.813	336	0.519	0.0845	0.0258	0.0869	0.0265	0.1391	0.0424
600	—	—	61	2.52	0.099	22.68	0.893	404	0.626	0.0704	0.0214	0.0732	0.0223	0.1159	0.0353

700	—	2.72	0.107	24.49	0.964	471	0.730	0.0603	0.0184	0.0622	0.0189	0.0994	0.0303
750	—	2.82	0.111	25.35	0.998	505	0.782	0.0563	0.0171	0.0579	0.0176	0.0927	0.0282
800	—	2.91	0.114	26.16	1.030	538	0.834	0.0528	0.0161	0.0544	0.0166	0.0868	0.0265
900	—	3.09	0.122	27.79	1.094	606	0.940	0.0470	0.0143	0.0481	0.0147	0.0770	0.0235
1000	—	3.25	0.128	29.26	1.152	673	1.042	0.0423	0.0129	0.0434	0.0132	0.0695	0.0212
1250	—	2.98	0.117	32.74	1.289	842	1.305	0.0338	0.0103	0.0347	0.0106	0.0554	0.0169
1500	—	3.26	0.128	35.86	1.412	1011	1.566	0.02814	0.00858	0.02814	0.00883	0.0464	0.0141
1750	—	2.98	0.117	38.76	1.526	1180	1.829	0.02410	0.00735	0.02410	0.00756	0.0397	0.0121
2000	—	3.19	0.126	41.45	1.632	1349	2.092	0.02109	0.00643	0.02109	0.00662	0.0348	0.0106

Notes:

1. These resistance values are valid **only** for the parameters as given. Using conductors having coated strands, different stranding type, and, especially, other temperatures changes the resistance.

2. Formula for temperature change: $P_2 = R_1 [1 + \alpha \cdot (T_2 - 75)]$ where: $\alpha_{CU} = 0.00323$, $\alpha_{AL} = 0.00330$ at 75°C.

3. Conductors with compact and compressed stranding have about 9 percent and 3 percent, respectively, smaller bare conductor diameters than those shown. See Table 5A of the *NEC* for actual compact cable dimensions.

4. The IACS conductivities used: bare copper = 100%, aluminum = 61%.

5. Class B stranding is listed as well as solid for some sizes. Its overall diameter and area is that of its circumscribing circle.

FPN: The construction information is per NEMA WC8-1992 or ANSI/UL 1581-1998. The resistance is calculated per National Bureau of Standards Handbook 100, dated 1966, and Handbook 109, dated 1972.

Source: NFPA 70, *National Electrical Code®*, NFPA, Quincy, MA, 2002 edition, Table 8.

CONDUCTORS

Exhibit VII.14 Conductor Application and Insulations

Trade Name	Type Letter	Maximum Operating Temperature	Application Provisions	Insulation	AWG or kcmil	mm	Mils	Outer Covering[1]
Fluorinated ethylene propylene	FEP or FEPB	90°C 194°F	Dry and damp locations	Fluorinated ethylene propylene	14–10 8–2	0.51 0.76	20 30	None
					14–8	0.36	14	Glass braid
		200°C 392°F	Dry locations— special applications[2]	Fluorinated ethylene propylene	6–2	0.36	14	Glass or other suitable braid material
Mineral insulation (metal sheathed)	MI	90°C 194°F	Dry and wet locations	Magnesium oxide	18–16[3] 16–10 9–4 3–500	0.58 0.91 1.27 1.40	23 36 50 55	Copper or alloy steel
		250°C 482°F	For special applications[2]					

Trade name	Type	Temp.	Application provisions	Insulation	AWG/kcmil	(A)	(B)	(A)	(B)	Outer covering
Moisture-, heat-, and oil-resistant thermoplastic	MTW	60°C 140°F	Machine tool wiring in wet locations.	Flame-retardant moisture-, heat-, and oil-resistant thermoplastic	22–12	0.76	0.38	30	15	(A) None (B) Nylon jacket or equivalent
		90°C 194°F	Machine tool wiring in dry locations.		10	0.76	0.51	30	20	
					8	1.14	0.76	45	30	
					6	1.52	0.76	60	30	
					4–2	1.52	1.02	60	40	
					1–4/0	2.03	1.27	80	50	
					213–500	2.41	1.52	95	60	
					501–1000	2.79	1.78	110	70	
Paper		85°C 185°F	For underground service conductors, or by special permission	Paper						Lead sheath
Perfluoroalkoxy	PFA	90°C 194°F	Dry and damp locations	Perfluoroalkoxy	14–10	0.51		20		None
		200°C 392°F	Dry locations—special applications[2]		8–2	0.76		30		
					1–4/0	1.14		45		

Continued

Exhibit VII.14 Conductor Application and Insulations *Continued*

Trade Name	Type Letter	Maximum Operating Temperature	Application Provisions	Insulation	AWG or kcmil	Thickness of Insulation mm	Thickness of Insulation Mils	Outer Covering[1]
Perfluoroalkoxy	PFAH	250°C 482°F	Dry locations only. Only for leads within apparatus or within raceways connected to apparatus (nickel or nickel-coated copper only)	Perfluoroalkoxy	14–10 8–2 1–4/0	0.51 0.76 1.14	20 30 45	None
Thermoset	RHH	90°C 194°F	Dry and damp locations		14–10 8–2 1–4/0 213–500 501–1000 1001–2000 For 601–2000 (see Table 310.62)	1.14 1.52 2.03 2.41 2.79 3.18	45 60 80 95 110 125	Moisture-resistant, flame-retardant, nonmetallic covering[1]

Trade Name	Type Letter	Maximum Operating Temperature	Application Provisions	Insulation	Size (AWG or kcmil)	Thickness of Insulation (mm)		Outer Covering
Moisture-resistant thermoset	RHW[4]	75°C 167°F	Dry and wet locations	Flame-retardant, moisture-resistant thermoset	14–10 8–2 1–4/0 213–500 501–1000 1001–2000 For 601–2000 (see Table 310.62)	1.14 1.52 2.03 2.41 2.79 3.18	45 60 80 95 110 125	Moisture-resistant, flame-retardant, nonmetallic covering[5]
Moisture-resistant thermoset	RHW-2	90°C 194°F	Dry and wet locations	Flame-retardant, moisture-resistant thermoset	14–10 8–2 1–4/0 213–500 501–1000 1001–2000 For 601–2000 (see Table 310.62)	1.14 1.52 2.03 2.41 2.79 3.18	45 60 80 95 110 125	Moisture-resistant, flame-retardant, nonmetallic covering[5]

Continued

Exhibit VII.14 Conductor Application and Insulations *Continued*

Trade Name	Type Letter	Maximum Operating Temperature	Application Provisions	Insulation	AWG or kcmil	mm	Mils	Outer Covering[1]
Silicone	SA	90°C 194°F	Dry and damp locations	Silicone rubber	14–10	1.14	45	Glass or other suitable braid material
					8–2	1.52	60	
		200°C 392°F	For special application[2]		1–4/0	2.03	80	
					213–500	2.41	95	
					501–1000	3.18	110	
					1001–2000	3.18	125	
Thermoset	SIS	90°C 194°F	Switchboard wiring only	Flame-retardant thermoset	14–10	0.76	30	None
					8–2	1.14	45	
					1–4/0	2.41	95	
Thermoplastic and fibrous outer braid	TBS	90°C 194°F	Switchboard wiring only	Thermoplastic	14–10	0.76	30	Flame-retardant, nonmetallic covering
					8	1.14	45	
					6–2	1.52	60	
					1–4/0	2.03	80	

	Type	Max. temp.	Application provisions	Insulation	AWG or kcmil	mm	mils	Outer covering
Extended polytetrafluoroethylene	TFE	250°C 482°F	Dry locations only. Only for leads within apparatus or within raceways connected to apparatus, or as open wiring (nickel or nickel-coated copper only)	Extruded polytetrafluoroethylene	14–10 8–2 1–4/0	0.51 0.76 1.14	20 30 45	None
Heat-resistant thermoplastic	THHN	90°C 194°F	Dry and damp locations	Flame-retardant, heat-resistant thermoplastic	14–12 10 8–6 4–2 1–4/0 250–500 501–1000	0.38 0.51 0.76 1.02 1.27 1.52 1.78	15 20 30 40 50 60 70	Nylon jacket or equivalent
Moisture- and heat-resistant thermoplastic	THHW	75°C 167°F 90°C 194°F	Wet location Dry location	Flame-retardant, moisture- and heat-resistant thermoplastic	14–10 8 6–2 1–4/0 213–500 501–1000	0.76 1.14 1.52 2.03 2.41 2.79	30 45 60 80 95 110	None

Continued

Exhibit VII.14 Conductor Application and Insulations *Continued*

Trade Name	Type Letter	Maximum Operating Temperature	Application Provisions	Insulation	Thickness of Insulation			Outer Covering[1]
					AWG or kcmil	mm	Mils	
Moisture- and heat-resistant thermoplastic	THW[4]	75°C 167°F	Dry and wet locations	Flame-retardant, moisture- and heat-resistant thermoplastic	14–10	0.76	30	None
					8	1.14	45	
					6–2	1.52	60	
		90°C 194°F	Special applications within electric discharge lighting equipment. Limited to 1000 open-circuit volts or less. (size 14–8 only)		1–4/0	2.03	80	
					213–500	2.41	95	
					501–1000	2.79	110	
					1001–2000	3.18	125	
Moisture- and heat-resistant thermoplastic	THWN[4]	75°C 167°F	Dry and wet locations	Flame-retardant, moisture- and heat-resistant thermoplastic	14–12	0.38	15	Nylon jacket or equivalent
					10	0.51	20	
					8–6	0.76	30	
					4–2	1.02	40	
					1–4/0	1.27	50	
					250–500	1.52	60	
					501–1000	1.78	70	

Trade name	Type	Max operating temperature	Application provisions	Insulation	Size AWG or kcmil	mm	mils	Outer covering
Moisture-resistant thermoplastic	TW	60°C 140°F	Dry and wet locations	Flame-retardant, moisture-resistant thermoplastic	14–10 8 6–2 1–4/0 213–500 501–1000 1001–2000	0.76 1.14 1.52 2.03 2.41 2.79 3.18	30 45 60 80 95 110 125	None
Underground feeder and branch-circuit cable—single conductor	UF	60°C 140°F 75°C 167°F[7]		Moisture-resistant Moisture- and heat-resistant	14–10 8–2 1–4/0	1.52 2.03 2.41	60[6] 80[6] 95[6]	Integral with insulation
Underground service-entrance cable—single conductor	USE[4]	75°C 167°F		Heat- and moisture-resistant	14–10 8–2 1–4/0 213–500 501–1000 1001–2000	1.14 1.52 2.03 2.41 2.79 3.18	45 60 80 95[8] 110 125	Moisture-resistant nonmetallic covering (see 338.2)

Continued

Exhibit VII.14 Conductor Application and Insulations *Continued*

Trade Name	Type Letter	Maximum Operating Temperature	Application Provisions	Insulation	AWG or kcmil	Thickness of Insulation		Outer Covering[1]
						mm	Mils	
Thermoset	XHH	90°C 194°F	Dry and damp locations	Flame-retardant thermoset	14–10	0.76	30	None
					8–2	1.14	45	
					1–4/0	1.40	55	
					213–500	1.65	65	
					501–1000	2.03	80	
					1001–2000	2.41	95	
Moisture-resistant thermoset	XHHW[4]	90°C 194°F	Dry and damp locations	Flame-retardant, moisture-resistant thermoset	14–10	0.76	30	None
		75°C 167°F	Wet locations		8–2	1.14	45	
					1–4/0	1.40	55	
					213–500	1.65	65	
					501–1000	2.03	80	
					1001–2000	2.41	95	

Continued

Trade Name	Type Letter	Max Operating Temperature	Application Provisions	Insulation	Size (AWG or kcmil)	Thickness (mm)	Thickness (mils)	Outer Covering
Moisture-resistant thermoset	XHHW-2	90°C 194°F	Dry and wet locations	Flame-retardant, moisture-resistant thermoset	14–10 8–2 1–4/0 213–500 501–1000 1001–2000	0.76 1.14 1.40 1.65 2.03 2.41	30 45 55 65 80 95	None
Modified ethylene tetrafluoro-ethylene	Z	90°C 194°F 150°C 302°F	Dry and damp locations Dry locations— special applications[2]	Modified ethylene tetrafluoroethylene	14–12 10 8–4 3–1 1/0–4/0	0.38 0.51 0.64 0.89 1.14	15 20 25 35 45	None
Modified ethylene tetrafluoro-ethylene	ZW[4]	75°C 167°F 90°C 194°F 150°C 302°F	Wet locations Dry and damp locations Dry locations— special applications[2]	Modified ethylene tetrafluoroethylene	14–10 8–2	0.76 1.14	30 45	None

Exhibit VII.14 Conductor Application and Insulations *Continued*

[1]Some insulations do not require an outer covering.

[2]Where design conditions require maximum conductor operating temperatures above 90°C (194°F).

[3]For signaling circuits permitting 300-V insulation.

[4]Listed wire types designated with the suffix "2," such as RHW-2, shall be permitted to be used at a continuous 90°C (194°F) operating temperature, wet or dry.

[5]Some rubber insulations do not require an outer covering.

[6]Includes integral jacket.

[7]For ampacity limitation, see 340.80 the *NEC*.

[8]Insulation thickness shall be permitted to be 2.03 mm (80 mils) for listed Type USE conductors that have been subjected to special investigations. The nonmetallic covering over individual rubber-covered conductors of aluminum-sheathed cable and of lead-sheathed or multiconductor cable shall not be required to be flame retardant.

Source: Adapted from NFPA 70, *National Electrical Code®*, NFPA, Quincy, MA, 2002 edition, Table 310.13.

Exhibit VII.15 Dimensions of Insulated Conductors and Fixture Wires

Type: FFH-2, RFH-1, RFH-2, RHH,ᵃ RHW,ᵃ RHW-2,ᵃ RHH, RHW, RHW-2, SF-1, SF-2, SFF-1, SFF-2, TF, TFF, THHW, THW, THW-2, TW, XF, XFF

Type	Size (AWG or kcmil)	Approximate Diameter		Approximate Area	
		mm	in.	mm²	in.²
RFH-2, FFH-2	18	3.454	0.136	9.355	0.0145
	16	3.759	0.148	11.10	0.0172
RHW-2, RHH, RHW	14	4.902	0.193	18.90	0.0293
	12	5.385	0.212	22.77	0.0353
	10	5.994	0.236	28.19	0.0437
	8	8.280	0.326	53.87	0.0835
	6	9.246	0.364	67.16	0.1041
	4	10.46	0.412	86.00	0.1333
	3	11.18	0.440	98.13	0.1521
	2	11.99	0.472	112.9	0.1750
	1	14.78	0.582	171.6	0.2660
	1/0	15.80	0.622	196.1	0.3039
	2/0	16.97	0.668	226.1	0.3505

Continued

Exhibit VII.15 Dimensions of Insulated Conductors and Fixture Wires *Continued*

Type: FFH-2, RFH-1, RFH-2, RHH, ª RHW, ª RHH, RHW, RHW-2, ª RHH, RHW, RHW-2, SF-1, SF-2, SFF-1, SFF-2, TF, TFF, THHW, THW, THW-2, TW, XF, XFF

Type	Size (AWG or kcmil)	Approximate Diameter		Approximate Area	
		mm	in.	mm²	in.²
RHW-2, RHH, RHW	3/0	18.29	0.720	262.7	0.4072
	4/0	19.76	0.778	306.7	0.4754
	250	22.73	0.895	405.9	0.6291
	300	24.13	0.950	457.3	0.7088
	350	25.43	1.001	507.7	0.7870
	400	26.62	1.048	556.5	0.8626
	500	28.78	1.133	650.5	1.0082
	600	31.57	1.243	782.9	1.2135
	700	33.38	1.314	874.9	1.3561
	750	34.24	1.348	920.8	1.4272
	800	35.05	1.380	965.0	1.4957
	900	36.68	1.444	1057	1.6377
	1000	38.15	1.502	1143	1.7719

Type	Size					
	1250	43.92	1.729		1515	2.3479
	1500	47.04	1.852		1738	2.6938
	1750	49.94	1.966		1959	3.0357
	2000	52.63	2.072		2175	3.3719
SF-2, SFF-2	18	3.073	0.121		7.419	0.0115
	16	3.378	0.133		8.968	0.0139
	14	3.759	0.148		11.10	0.0172
SF-1, SFF-1	18	2.311	0.091		4.194	0.0065
RFH-1, XF, XFF	18	2.692	0.106		5.161	0.0080
TF, TFF, XF, XFF	16	2.997	0.118		7.032	0.0109
TW, XF, XFF, THHW, THW, THW-2	14	3.378	0.133		8.968	0.0139
TW, THHW, THW, THW-2	12	3.861	0.152		11.68	0.0181
	10	4.470	0.176		15.68	0.0243
	8	5.994	0.236		28.19	0.0437
RHH,[a] RHW,[a] RHW7-2[a]	14	4.140	0.163		13.48	0.0209
	12	4.623	0.182		16.77	0.0260

Continued

Exhibit VII.15 Dimensions of Insulated Conductors and Fixture Wires *Continued*

Type	Size (AWG or kcmil)	Approximate Diameter		Approximate Area	
		mm	in.	mm²	in.²
Type: RHH,ᵃ RHW,ᵃ RHW-2,ᵃ RHH,ᵃ THHN, THHW, THW, THW-2, TFN, TFFN, THWN, THWN-2, XF, XFF					
THHW, THW, AF, XF, XFF	10	5.232	0.206	21.48	0.0333
RHH,ᵃ RHW,ᵃ RHW-2ᵃ	8	6.756	0.266	35.87	0.0556
TW, THW, THHW, THW-2, RHH,ᵃ RHW,ᵃ RHW-2ᵃ	6	7.722	0.304	46.84	0.0726
	4	8.941	0.352	62.77	0.0973
	3	9.652	0.380	73.16	0.1134
	2	10.46	0.412	86.00	0.1333
	1	12.50	0.492	122.6	0.1901
	1/0	13.51	0.532	143.4	0.2223
	2/0	14.68	0.578	169.3	0.2624
	3/0	16.00	0.630	201.1	0.3117
	4/0	17.48	0.688	239.9	0.3718
	250	19.43	0.765	296.5	0.4596
	300	20.83	0.820	340.7	0.5281

350	22.12	0.871	384.4	0.5958
400	23.32	0.918	427.0	0.6619
500	25.48	1.003	509.7	0.7901
600	28.27	1.113	627.7	0.9729
700	30.07	1.184	710.3	1.1010
750	30.94	1.218	751.7	1.1652
800	31.75	1.250	791.7	1.2272
900	33.38	1.314	874.9	1.3561
1000	34.85	1.372	953.8	1.4784
1250	39.09	1.539	1200	1.8602
1500	42.21	1.662	1400	2.1695
1750	45.11	1.776	1598	2.4773
2000	47.80	1.882	1795	2.7818
TFN, TFFN				
18	2.134	0.084	3.548	0.0055
16	2.438	0.096	4.645	0.0072
THHN, THWN, THWN-2				
14	2.819	0.111	6.258	0.0097
12	3.302	0.130	8.581	0.0133
10	4.166	0.164	13.61	0.0211

Continued

Exhibit VII.15 Dimensions of Insulated Conductors and Fixture Wires *Continued*

Type	Size (AWG or kcmil)	Approximate Diameter		Approximate Area	
		mm	in.	mm²	in.²
Type: RHH, a RHW, a RHW-2, a THHN, THHW, THW, THW-2, TFN, TFFN, THWN, THWN-2, XF, XFF					
THHN, THWN, THWN-2	8	5.486	0.216	23.61	0.0366
	6	6.452	0.254	32.71	0.0507
	4	8.230	0.324	53.16	0.0824
	3	8.941	0.352	62.77	0.0973
	2	9.754	0.384	74.71	0.1158
	1	11.33	0.446	100.8	0.1562
	1/0	12.34	0.486	119.7	0.1855
	2/0	13.51	0.532	143.4	0.2223
	3/0	14.83	0.584	172.8	0.2679
	4/0	16.31	0.642	208.8	0.3237
	250	18.06	0.711	256.1	0.3970
	300	19.46	0.766	297.3	0.4608

Type: FEP, FEPB, PAF, PAFF, PF, PFA, PFAH, PFF, PGF, PGFF, PTF, PTFF, TFE, THHN, THWN, THWN-2, Z, ZF, ZEF

THHN, THWN, THWN-2				
350	20.75	0.817	338.2	0.5242
400	21.95	0.864	378.3	0.5863
500	24.10	0.949	456.3	0.7073
600	26.70	1.051	559.7	0.8676
700	28.50	1.122	637.9	0.9887
750	29.36	1.156	677.2	1.0496
800	30.18	1.188	715.2	1.1085
900	31.80	1.252	794.3	1.2311
1000	33.27	1.310	869.5	1.3478
PF, PGFF, PGF, PFF, PTF, PAF, PTFF, PAFF				
18	2.184	0.086	3.742	0.0058
16	2.489	0.098	4.839	0.0075
PF, PGFF, PGF, PFF, PTF, PAF, PTFF, PAFF, TFE, FEP, PFA, FEPB, PFAH				
14	2.870	0.113	6.452	0.0100
TFE, FEP, PFA, FEPB, PFAH				
12	3.353	0.132	8.839	0.0137
10	3.962	0.156	12.32	0.0191
8	5.232	0.206	21.48	0.0333
6	6.198	0.244	30.19	0.0468

Continued

Exhibit VII.15 Dimensions of Insulated Conductors and Fixture Wires *Continued*

Type	Size (AWG or kcmil)	Approximate Diameter		Approximate Area	
		mm	in.	mm²	in.²
Type: FEP, FEPB, PAF, PAFF, PF, PFA, PFAH, PFF, PGF, PGFF, PTF, PTFE, TFE, THHN, THWN, TWHN-2, Z, ZF, ZFF					
TFE, FEP, PFA, FEPB, PFAH	4	7.417	0.292	43.23	0.0670
	3	8.128	0.320	51.87	0.0804
	2	8.941	0.352	62.77	0.0973
TFE, PFAH	1	10.72	0.422	90.26	0.1399
TFE, PFA, PFAH, Z	1/0	11.73	0.462	108.1	0.1676
	2/0	12.90	0.508	130.8	0.2027
	3/0	14.22	0.560	158.9	0.2463
	4/0	15.70	0.618	193.5	0.3000
ZF, ZFF	18	1.930	0.076	2.903	0.0045
	16	2.235	0.088	3.935	0.0061
Z, ZF, ZFF	14	2.616	0.103	5.355	0.0083
Z	12	3.099	0.122	7.548	0.0117
	10	3.962	0.156	12.32	0.0191

	Size				
XHHW, ZW, XHHW-2, XHH	8	4.978	0.196	19.48	0.0302
	6	5.944	0.234	27.74	0.0430
	4	7.163	0.282	40.32	0.0625
	3	8.382	0.330	55.16	0.0855
	2	9.195	0.362	66.39	0.1029
	1	10.21	0.402	81.87	0.1269
Type: KF-1, KF-2, KFF-1, KFF-2, XHH, XHHW, XHHW-2, ZW					
	14	3.378	0.133	8.968	0.0139
	12	3.861	0.152	11.68	0.0181
	10	4.470	0.176	15.68	0.0243
	8	5.994	0.236	28.19	0.0437
	6	6.960	0.274	38.06	0.0590
	4	8.179	0.322	52.52	0.0814
	3	8.890	0.350	62.06	0.0962
	2	9.703	0.382	73.94	0.1146
XHHW, XHHW-2, XHH	1	11.23	0.442	98.97	0.1534
	1/0	12.24	0.482	117.7	0.1825
	2/0	13.41	0.528	141.3	0.2190

Continued

Exhibit VII.15 Dimensions of Insulated Conductors and Fixture Wires *Continued*

Type	Size (AWG or kcmil)	Approximate Diameter		Approximate Area	
		mm	in.	mm²	in.²
	Type: KF-1, KF-2, KFF-1, KFF-2, XHH, XHHW, XHHW-2, ZW				
XHHW, XHHW-2, XHH	3/0	14.73	0.580	170.5	0.2642
	4/0	16.21	0.638	206.3	0.3197
	250	17.91	0.705	251.9	0.3904
	300	19.30	0.760	292.6	0.4536
	350	20.60	0.811	333.3	0.5166
	400	21.79	0.858	373.0	0.5782
	500	23.95	0.943	450.6	0.6984
	600	26.75	1.053	561.9	0.8709
	700	28.55	1.124	640.2	0.9923
	750	29.41	1.158	679.5	1.0532
	800	30.23	1.190	717.5	1.1122
	900	31.85	1.254	796.8	1.2351
	1000	33.32	1.312	872.2	1.3519
	1250	37.57	1.479	1108	1.7180

1500	40.69	1.602	1300	2.0157
1750	43.59	1.716	1492	2.3127
2000	46.28	1.822	1682	2.6073
KF-2, KFF-2				
18	1.600	0.063	2.000	0.0031
16	1.905	0.075	2.839	0.0044
14	2.286	0.090	4.129	0.0064
12	2.769	0.109	6.000	0.0093
10	3.378	0.133	8.968	0.0139
KF-1, KFF-1				
18	1.448	0.057	1.677	0.0026
16	1.753	0.069	2.387	0.0037
14	2.134	0.084	3.548	0.0055
12	2.616	0.103	5.355	0.0083
10	3.226	0.127	8.194	0.0127

[a]Types RHH, RHW, and RHW-2 without outer covering.

Source: NFPA 70, *National Electrical Code®*, NFPA, Quincy, MA, 2002 edition, Table 5.

Exhibit VII.16 Remote Control and Signaling Cable Types and Uses

Cable Type	Description
CL2P and CL3P	Types CL2P and CL3P plenum cables are listed as being suitable for use in ducts, plenums, and other space used for environmental air and are also listed as having adequate fire-resistant and low smoke-producing characteristics.
CL2R and CL3R	Types CL2R and CL3R riser cables are listed as being suitable for use in a vertical run in a shaft or from floor to floor and are also listed as having fire-resistant characteristics capable of preventing the carrying of fire from floor to floor.
PLTC	Type PLTC nonmetallic-sheathed, power-limited tray cable is listed as being suitable for cable trays and consists of a factory assembly of two or more insulated conductors under a nonmetallic jacket. The insulated conductors are 22 AWG through 12 AWG. The conductor material is copper (solid or stranded). Insulation on conductors is suitable for 300 volts. The cable core is either (1) two or more parallel conductors, (2) one or more group assemblies of twisted or parallel conductors, or (3) a combination thereof. A metallic shield or a metallized foil shield with drain wire(s) is permitted to be applied either over the cable core, over groups of conductors, or both. The cable is listed as being resistant to the spread of fire. The outer jacket is sunlight- and moisture-resistant nonmetallic material.

CL2 and CL3	Types CL2 and CL3 cables are listed as being suitable for general-purpose use, with the exception of risers, ducts, plenums, and other space used for environmental air and are also listed as being resistant to the spread of fire.
CL2X and CL3X	Types CL2X and CL3X limited-use cables are listed as being suitable for use in dwellings and for use in raceway and are also listed as being resistant to flame spread.
Class 3 Single Conductors	Class 3 single conductors used as other wiring within buildings cannot be smaller than 18 AWG and must be Type CL3. Conductor types described in 725.27(B) of the *NEC* that are also listed as Type CL3 are permitted.

Source: Adapted from NFPA 70, *National Electrical Code®*, NFPA, Quincy, MA, 2002 edition, Section 725-71.

Exhibit VII.17 Cable Uses and Permitted Substitutions for Class 2, Class 3, and PLTC Cables

Cable Type	Use	References	Permitted Substitutions
CL3P	Class 3 plenum cable	725.61(A)	CMP
CL2P	Class 2 plenum cable	725.61(A)	CMP, CL3P
CL3R	Class 3 riser cable	725.61(B)	CMP, CL3P, CMR
CL2R	Class 2 riser cable	725.61(B)	CMP, CL3P, CL2P, CMR, CL3R
PLTC	Power-limited tray cable	725.61(C) and (D)	
CL3	Class 3 cable	725.61(B), (E), and (F)	CMP, CL3P, CMR, CL3R, CMG, CM, PLTC
CL2	Class 2 cable	725.61(B), (E), and (F)	CMP, CL3P, CL2P, CMR, CL3R, CL2R, CMG, CM, PLTC, CL3
CL3X	Class 3 cable, limited use	725.61(B) and (E)	CMP, CL3P, CMR, CL3R, CMG, CM, PLTC, CL3, CMX
CL2X	Class 2 cable, limited use	725.61(B) and (E)	CMP, CL3P, CL2P, CMR, CL3R, CL2R, CMG, CM, PLTC, CL3, CL2, CMX, CL3X

Source: NFPA 70, National Electrical Code®, NFPA, Quincy, MA, 2002 edition, Table 725.61.

CABLES

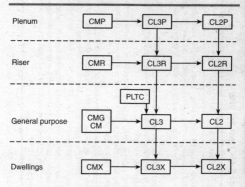

Type CM—Communications wires and cables

Type CL2 and CL3—Class 2 and Class 3 remote-control, signaling, and power-limited cables

Type PLTC—Power-limited tray cable

A ➔ B Cable A shall be permitted to be used in place of cable B.

Source: NFPA 70, National Electrical Code®, NFPA, Quincy, MA, 2002 edition, Fig. 725.61.

Exhibit VII.19 Non-Power-Limited Cable Types and Uses

Cable Type	Description
NPLFP	Type NPLFP non-power-limited fire alarm cable for use in other space used for environmental air is listed as being suitable for use in other space used for environmental air as described in 300.22(C) of the *NEC* and is also listed as having adequate fire-resistant and low smoke-producing characteristics.
NPLFR	Type NPLFR non-power-limited fire alarm riser cable is listed as being suitable for use in a vertical run in a shaft or from floor to floor and is also listed as having fire-resistant characteristics capable of preventing the carrying of fire from floor to floor.
NPLF	Type NPLF non-power-limited fire alarm cable is listed as being suitable for general-purpose fire alarm use, with the exception of risers, ducts, plenums, and other space used for environmental air, and is also listed as being resistant to the spread of fire.
Fire alarm circuit integrity (CI) cable	Cables suitable for use in fire alarm systems to ensure survivability of critical circuits during a specified time under fire conditions are listed as circuit integrity (CI) cable. Cables identified in 760.31(C), (D), and (E) of the *NEC* that meet the requirements for circuit integrity have the additional classification using the suffix "CI" (for example, NPLFP-CI, NPLFR-CI, and NPLF-CI).
NPLFA cable markings	Multiconductor non-power-limited fire alarm cables are marked in accordance with Table 760.31(G) of the *NEC*. Non-power-limited fire alarm circuit cables are permitted to be marked with a maximum usage voltage rating of 150 volts. Cables that are listed for circuit integrity are identified with the suffix "CI" as defined in 760.31(F).

Source: Adapted from NFPA 70, *National Electrical Code*®, NFPA, Quincy, MA, 2002 edition, Section 760-31.

Exhibit VII.20 Power-Limited Cable Types and Uses

Cable Type	Description
FPLP	Type FPLP power-limited fire alarm plenum cable is listed as being suitable for use in ducts, plenums, and other space used for environmental air and is also listed as having adequate fire-resistant and low smoke-producing characteristics.
FPLR	Type FPLR power-limited fire alarm riser cable is listed as being suitable for use in a vertical run in a shaft or from floor to floor and is also listed as having fire-resistant characteristics capable of preventing the carrying of fire from floor to floor.
FPL	Type FPL power-limited fire alarm cable is listed as being suitable for general-purpose fire alarm use, with the exception of risers, ducts, plenums, and other spaces used for environmental air and is also listed as being resistant to the spread of fire.
Fire alarm circuit integrity (CI) cable	Cables suitable for use in fire alarm systems to ensure survivability of critical circuits during a specified time under fire conditions are listed as circuit integrity (CI) cable. Cables identified in 760.71(D), (E), and (F) of the *NEC* that meet the requirements for circuit integrity have the additional classification using the suffix "CI" (for example, FPLP-CI, FPLR-CI, and FPL-CI).
Coaxial cables	Coaxial cables are permitted to use 30 percent conductivity copper-covered steel center conductor wire and are listed as Type FPLP, FPLR, or FPL cable.

Source: Adapted from NFPA 70, *National Electrical Code®*, NFPA, Quincy, MA, 2002 edition, Section 760-71.

Exhibit VII.21 Cable Uses and Permitted Substitutions for Power-Limited Fire Alarm Cables

| Cable Type | Use | References | Permitted Substitutions | |
			Multiconductor	Coaxial
FPLP	Power-limited fire alarm plenum cable	760.61(A)	CMP	MPP
FPLR	Power-limited fire alarm riser cable	760.61(B)	CMP, FPLP, CMR	MPP, MPR
FPL	Power-limited fire alarm cable	760.61(C)	CMP, FPLP, CMR, FPLR, CMG, CM	MPP, MPR, MPG, MP

Source: NFPA 70, *National Electrical Code®*, NFPA, Quincy, MA, 2002 edition; Table 760.61.

CABLES

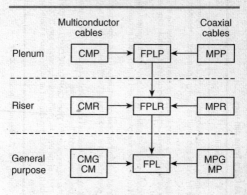

Exhibit VII.23 Communications Cable Types and Uses

Cable Type	Description
CMP	Type CMP communications plenum cable is listed as being suitable for use in ducts, plenums, and other spaces used for environmental air and is also listed as having adequate fire-resistant and low smoke-producing characteristics.
CMR	Type CMR communications riser cable is listed as being suitable for use in a vertical run in a shaft or from floor to floor and is also listed as having fire-resistant characteristics capable of preventing the carrying of fire from floor to floor.
CMG	Type CMG general-purpose communications cable is listed as being suitable for general-purpose communications use, with the exception of risers and plenums, and is also listed as being resistant to the spread of fire.
CM	Type CM communications cable is listed as being suitable for general-purpose communications use, with the exception of risers and plenums, and is also listed as being resistant to the spread of fire.
CMX	Type CMX limited-use communications cable is listed as being suitable for use in dwellings and for use in raceway and is also listed as being resistant to flame spread.
Multipurpose (MP) cables	Until July 1, 2003, cables that meet the requirements for Types CMP, CMR, CMG, and CM and also satisfy the requirements of 760.71(B) of the *NEC* for multiconductor cables and 760.71(H) of the *NEC* for coaxial cables are permitted to be listed and marked as multipurpose cable Types MPP, MPR, MPG, and MP, respectively.

Source: Adapted from NFPA 70, *National Electrical Code®*, NFPA, Quincy, MA, 2002 edition, Section 800-51.

Exhibit VII.24 Cable Uses and Permitted Substitutions for Communications Cables

Cable Type	Use	References	Permitted Substitutions
CMP	Communications plenum cable	800.53(A)	MPP
CMR	Communications riser cable	800.53(B)	MPP, CMP, MPR
CMG, CM	Communications general-purpose cable	800.53(E)(1)	MPP, CMP, MPR, CMR, MPG, MP
CMX	Communications cable, limited use	800.53(E)	MPP, CMP, MPR, CMR, MPG, MP, CMG, CM

Note: See Exhibit VII.25, Cable substitution hierarchy.
Source: NFPA 70, *National Electrical Code®*, NFPA, Quincy, MA, 2002 edition, Table 800.53.

CABLES

Exhibit VII.25 Cable Substitution Hierarchy for Communications Cables

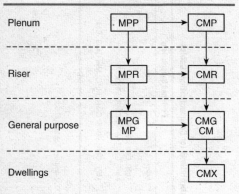

Type CM—Communications cables

Type MP—Multipurpose cable

A → B   Cable A shall be permitted to be used in place of cable B.

Source: NFPA 70, *National Electrical Code®*, NFPA, Quincy, MA, 2002 edition, Fig. 800.53.

CONDUIT AND RACEWAY SYSTEMS

**Exhibit VII.26 Percent of Cross Section
of Conduit and Tubing for Conductors**

Number of Conductors	All Conductor Types
1	53
2	31
Over 2	40

*Source: NFPA 70, National Electrical Code®, NFPA, Quincy, MA,
2002 edition, Table 1.*

*Editor's Note: This table is based on common conditions of proper
cabling and alignment of conductors where the length of the pull
and the number of bends are within reasonable limits. It should be
recognized that, for certain conditions, a larger size conduit or a
lesser conduit fill should be considered. When pulling three conductors or cables into a raceway, if the ratio of the raceway (inside
diameter) to the conductor or cable (outside diameter) is between
2.8 and 3.2, jamming can occur. While jamming can occur when
pulling four or more conductors or cables into a raceway, the prob-
ability is very low.*

Exhibit VII.27 Dimensions and Percent Area of Electrical Metallic Tubing (EMT)

Metric Designator	Trade Size	Nominal Internal Diameter		Total Area 100%		2 Wires 31%		Over 2 Wires 40%		1 Wire 53%		60%	
		mm	in.	mm²	in.²	mm²	in.²	mm²	in.²	mm²	in.²	mm²	in.²
16	1/2	15.8	0.622	196	0.304	61	0.094	78	0.122	104	0.161	118	0.182
21	3/4	20.9	0.824	343	0.533	106	0.165	137	0.213	182	0.283	206	0.320
27	1	26.6	1.049	556	0.864	172	0.268	222	0.346	295	0.458	333	0.519
35	1-1/4	35.1	1.380	968	1.496	300	0.464	387	0.598	513	0.793	581	0.897
41	1-1/2	40.9	1.610	1314	2.036	407	0.631	526	0.814	696	1.079	788	1.221
53	2	52.5	2.067	2165	3.356	671	1.040	866	1.342	1147	1.778	1299	2.013
63	2-1/2	69.4	2.731	3783	5.888	1173	1.816	1513	2.343	2005	3.105	2270	3.515
78	3	85.2	3.356	5701	8.846	1767	2.742	2280	3.538	3022	4.688	3421	5.307
91	3-1/2	97.4	3.834	7451	11.545	2310	3.579	2980	4.618	3949	6.119	4471	6.927
103	4	110.1	4.334	9521	14.753	2951	4.573	3808	5.901	5046	7.819	5712	8.852

Note: Refer to the *NEC* for more information on conduit fill.

Source: NFPA 70, *National Electrical Code®*, NFPA, Quincy, MA, 2002 edition, Table 4.

Exhibit VII.28 Dimensions and Percent Area of Electrical Nonmetallic Tubing (ENT)

Metric Designator	Trade Size	Nominal Internal Diameter		Total Area 100%		2 Wires 31%		Over 2 Wires 40%		1 Wire 53%		60%	
		mm	in.	mm²	in.²	mm²	in.²	mm²	in.²	mm²	in.²	mm²	in.²
16	1/2	14.2	0.560	158	0.246	49	0.076	63	0.099	84	0.131	95	0.148
21	3/4	19.3	0.760	293	0.454	91	0.141	117	0.181	155	0.240	176	0.272
27	1	25.4	1.000	507	0.785	157	0.243	203	0.314	269	0.416	304	0.471
35	1-1/4	34.0	1.340	908	1.410	281	0.437	363	0.564	481	0.747	545	0.846
41	1-1/2	39.9	1.570	1250	1.936	388	0.600	500	0.774	663	1.026	750	1.162
53	2	51.3	2.020	2067	3.205	641	0.993	827	1.282	1095	1.699	1240	1.923
63	2-1/2	—	—	—	—	—	—	—	—	—	—	—	—
78	3	—	—	—	—	—	—	—	—	—	—	—	—
91	3-1/2	—	—	—	—	—	—	—	—	—	—	—	—

Note: Refer to the NEC for more information on conduit fill.

Source: NFPA 70, National Electrical Code®, NFPA, Quincy, MA, 2002 edition, Table 4.

Exhibit VII.29 Dimensions and Percent Area of Flexible Metal Conduit (FMC)

Metric Designator	Trade Size	Nominal Internal Diameter		Total Area 100%		2 Wires 31%		Over 2 Wires 40%		1 Wire 53%		60%	
		mm	in.	mm²	in.²	mm²	in.²	mm²	in.²	mm²	in.²	mm²	in.²
12	3/8	9.7	0.384	74	0.116	23	0.036	30	0.046	39	0.061	44	0.069
16	1/2	16.1	0.635	204	0.317	63	0.098	81	0.127	108	0.168	122	0.190
21	3/4	20.9	0.824	343	0.533	106	0.165	137	0.213	182	0.283	206	0.320
27	1	25.9	1.020	527	0.817	163	0.253	211	0.327	279	0.433	316	0.490
35	1-1/4	32.4	1.275	824	1.277	256	0.396	330	0.511	437	0.677	495	0.766
41	1-1/2	39.1	1.538	1201	1.858	372	0.576	480	0.743	636	0.985	720	1.115
53	2	51.8	2.040	2107	3.269	653	1.013	843	1.307	1117	1.732	1264	1.961
63	2-1/2	63.5	2.500	3167	4.909	982	1.522	1267	1.963	1678	2.602	1900	2.945
78	3	76.2	3.000	4560	7.069	1414	2.191	1824	2.827	2417	3.746	2736	4.241
91	3-1/2	88.9	3.500	6207	9.621	1924	2.983	2483	3.848	3290	5.099	3724	5.773
103	4	101.6	4.000	8107	12.566	2513	3.896	3243	5.027	4297	6.660	4864	7.540

Note: Refer to the *NEC* for more information on conduit fill.

Source: NFPA 70, *National Electrical Code®*, NFPA, Quincy, MA, 2002 edition, Table 4.

Exhibit VII.30 Dimensions and Percent Area of Intermediate Metal Conduit (IMC)

Metric Designator	Trade Size	Nominal Internal Diameter		Total Area 100%		2 Wires 31%		Over 2 Wires 40%		1 Wire 53%		60%	
		mm	in.	mm²	in.²	mm²	in.²	mm²	in.²	mm²	in.²	mm²	in.²
12	3/8	—	—	—	—	—	—	—	—	—	—	—	—
16	1/2	16.8	0.660	222	0.342	69	0.106	89	0.137	117	0.181	133	0.205
21	3/4	21.9	0.864	377	0.586	117	0.182	151	0.235	200	0.311	226	0.352
27	1	28.1	1.105	620	0.959	192	0.297	248	0.384	329	0.508	372	0.575
35	1-1/4	36.8	1.448	1064	1.647	330	0.510	425	0.659	564	0.873	638	0.988
41	1-1/2	42.7	1.683	1432	2.225	444	0.690	573	0.890	759	1.179	859	1.335
53	2	54.6	2.150	2341	3.630	726	1.125	937	1.452	1241	1.924	1405	2.178
63	2-1/2	64.9	2.557	3308	5.135	1026	1.592	1323	2.054	1753	2.722	1985	3.081
78	3	80.7	3.176	5115	7.922	1586	2.456	2046	3.169	2711	4.199	3069	4.753
91	3-1/2	93.2	3.671	6822	10.584	2115	3.281	2729	4.234	3616	5.610	4093	6.351
103	4	105.4	4.166	8725	13.631	2705	4.226	3490	5.452	4624	7.224	5235	8.179

Note: Refer to the *NEC* for more information on conduit fill.

Source: NFPA 70, *National Electrical Code*®, NFPA, Quincy, MA, 2002 edition, Table 4.

Exhibit VII.31 Dimensions and Percent Area of Liquidtight Flexible Nonmetallic Conduit (LFNC-B[a])

Metric Designator	Trade Size	Nominal Internal Diameter		Total Area 100%		2 Wires 31%		Over 2 Wires 40%		1 Wire 53%		60%	
		mm	in.	mm²	in.²	mm²	in.²	mm²	in.²	mm²	in.²	mm²	in.²
12	3/8	12.5	0.494	123	0.192	38	0.059	49	0.077	65	0.102	74	0.115
16	1/2	16.1	0.632	204	0.314	63	0.097	81	0.125	108	0.166	122	0.188
21	3/4	21.1	0.830	350	0.541	108	0.168	140	0.216	185	0.287	210	0.325
27	1	26.8	1.054	564	0.873	175	0.270	226	0.349	299	0.462	338	0.524
35	1-1/4	35.4	1.395	984	1.528	305	0.474	394	0.611	522	0.810	591	0.917
41	1-1/2	40.3	1.588	1276	1.981	395	0.614	510	0.792	676	1.050	765	1.188
53	2	51.6	2.033	2091	3.246	648	1.006	836	1.298	1108	1.720	1255	1.948

[a]Corresponds to 356.2(2) of the NEC.

Note: Refer to the NEC for more information on conduit fill.

Source: NFPA 70, National Electrical Code®, NFPA, Quincy, MA, 2002 edition, Table 4.

Exhibit VII.32 Dimensions and Percent Area of Liquidtight Flexible Nonmetallic Conduit (LFNC-A[a])

Metric Designator	Trade Size	Nominal Internal Diameter		Total Area 100%		2 Wires 31%		Over 2 Wires 40%		1 Wire 53%		60%	
		mm	in.	mm²	in.²	mm²	in.²	mm²	in.²	mm²	in.²	mm²	in.²
12	3/8	12.6	0.495	125	0.192	39	0.060	50	0.077	66	0.102	75	0.115
16	1/2	16.0	0.630	201	0.312	62	0.097	80	0.125	107	0.165	121	0.187
21	3/4	21.0	0.825	346	0.535	107	0.166	139	0.214	184	0.283	208	0.321
27	1	26.5	1.043	552	0.854	171	0.265	221	0.342	292	0.453	331	0.513
35	1-1/4	35.1	1.383	968	1.502	300	0.466	387	0.601	513	0.796	581	0.901
41	1-1/2	40.7	1.603	1301	2.018	403	0.626	520	0.807	690	1.070	781	1.211
53	2	52.4	2.063	2157	3.343	669	1.036	863	1.337	1143	1.772	1294	2.006

[a]Corresponds to 356.2(1) of the NEC.

Note: Refer to the NEC for more information on conduit fill.

Source: NFPA 70, National Electrical Code®, NFPA, Quincy, MA, 2002 edition, Table 4.

Exhibit VII.33 Dimensions and Percent Area of Liquidtight Flexible Metal Conduit (LFMC)

Metric Designator	Trade Size	Nominal Internal Diameter		Total Area 100%		2 Wires 31%		Over 2 Wires 40%		1 Wire 53%		60%	
		mm	in.	mm²	in.²	mm²	in.²	mm²	in.²	mm²	in.²	mm²	in.²
12	3/8	12.5	0.494	123	0.192	38	0.059	49	0.077	65	0.102	74	0.115
16	1/2	16.1	0.632	204	0.314	63	0.097	81	0.125	108	0.166	122	0.188
21	3/4	21.1	0.830	350	0.541	108	0.168	140	0.216	185	0.287	210	0.325
27	1	26.8	1.054	564	0.873	175	0.270	226	0.349	299	0.462	338	0.524
35	1-1/4	35.4	1.395	984	1.528	305	0.474	394	0.611	522	0.810	591	0.917
41	1-1/2	40.3	1.588	1276	1.981	395	0.614	510	0.792	676	1.050	765	1.188
53	2	51.6	2.033	2091	3.246	648	1.006	836	1.298	1108	1.720	1255	1.948
63	2-1/2	63.3	2.493	3147	4.881	976	1.513	1259	1.953	1668	2.587	1888	2.929
78	3	78.4	3.085	4827	7.475	1497	2.317	1931	2.990	2559	3.962	2896	4.485
91	3-1/2	89.4	3.520	6277	9.731	1946	3.017	2511	3.893	3327	5.158	3766	5.839
103	4	102.1	4.020	8187	12.692	2538	3.935	3275	5.077	4339	6.727	4912	7.615
129	5	—	—	—	—	—	—	—	—	—	—	—	—
155	6	—	—	—	—	—	—	—	—	—	—	—	—

Note: Refer to the *NEC* for more information on conduit fill.

Source: NFPA 70, *National Electrical Code®*, NFPA, Quincy, MA, 2002 edition, Table 4.

Exhibit VII.34 Dimensions and Percent Area of Rigid Metal Conduit (RMC)

Metric Designator	Trade Size	Nominal Internal Diameter		Total Area 100%		2 Wires 31%		Over 2 Wires 40%		1 Wire 53%		60%	
		mm	in.	mm²	in.²	mm²	in.²	mm²	in.²	mm²	in.²	mm²	in.²
12	3/8												
16	1/2	16.1	0.632	204	0.314	63	0.097	81	0.125	108	0.166	122	0.188
21	3/4	21.2	0.836	353	0.549	109	0.170	141	0.220	187	0.291	212	0.329
27	1	27.0	1.063	573	0.887	177	0.275	229	0.355	303	0.470	344	0.532
35	1-1/4	35.4	1.394	984	1.526	305	0.473	394	0.610	522	0.809	591	0.916
41	1-1/2	41.2	1.624	1333	2.071	413	0.642	533	0.829	707	1.098	800	1.243
53	2	52.9	2.083	2198	3.408	681	1.056	879	1.363	1165	1.806	1319	2.045
63	2-1/2	63.2	2.489	3137	4.866	972	1.508	1255	1.946	1663	2.579	1882	2.919
78	3	78.5	3.090	4840	7.499	1500	2.325	1936	3.000	2565	3.974	2904	4.499
91	3-1/2	90.7	3.570	6461	10.010	2003	3.103	2584	4.004	3424	5.305	3877	6.006
103	4	102.9	4.050	8316	12.882	2578	3.994	3326	5.153	4408	6.828	4990	7.729
129	5	128.9	5.073	13050	20.212	4045	6.266	5220	8.085	6916	10.713	7830	12.127
155	6	154.8	6.093	18821	29.158	5834	9.039	7528	11.663	9975	15.454	11292	17.495

Note: Refer to the *NEC* for more information on conduit fill.
Source: NFPA 70, *National Electrical Code®*, NFPA, Quincy, MA, 2002 edition, Table 4.

Exhibit VII.35 Dimensions and Percent Area of Rigid PVC Conduit (RNC), Schedule 80

Metric Designator	Trade Size	Nominal Internal Diameter		Total Area 100%		2 Wires 31%		Over 2 Wires 40%		1 Wire 53%		60%	
		mm	in.	mm²	in.²	mm²	in.²	mm²	in.²	mm²	in.²	mm²	in.²
12	3/8	—	—	—	—	—	—	—	—	—	—	—	—
16	1/2	13.4	0.526	141	0.217	44	0.067	56	0.087	75	0.115	85	0.130
21	3/4	18.3	0.722	263	0.409	82	0.127	105	0.164	139	0.217	158	0.246
27	1	23.8	0.936	445	0.688	138	0.213	178	0.275	236	0.365	267	0.413
35	1-1/4	31.9	1.255	799	1.237	248	0.383	320	0.495	424	0.656	480	0.742
41	1-1/2	37.5	1.476	1104	1.711	342	0.530	442	0.684	585	0.907	663	1.027
53	2	48.6	1.913	1855	2.874	575	0.891	742	1.150	983	1.523	1113	1.725
63	2-1/2	58.2	2.290	2660	4.119	825	1.277	1064	1.647	1410	2.183	1596	2.471
78	3	72.7	2.864	4151	6.442	1287	1.997	1660	2.577	2200	3.414	2491	3.865
91	3-1/2	84.5	3.326	5608	8.688	1738	2.693	2243	3.475	2972	4.605	3365	5.213
103	4	96.2	3.786	7268	11.258	2253	3.490	2907	4.503	3852	5.967	4361	6.755
129	5	121.1	4.768	11518	17.855	3571	5.535	4607	7.142	6105	9.463	6911	10.713
155	6	145.0	5.709	16513	25.598	5119	7.935	6605	10.239	8752	13.567	9908	15.359

Note: Refer to the *NEC* for more information on conduit fill.
Source: NFPA 70, *National Electrical Code®*, NFPA, Quincy, MA, 2002 edition, Table 4.

Exhibit VII.36 Dimensions and Percent Area of Rigid PVC Conduit (RNC), Schedule 40, and HDPE Conduit

Metric Designator	Trade Size	Nominal Internal Diameter mm	Nominal Internal Diameter in.	Total Area 100% mm²	Total Area 100% in.²	2 Wires 31% mm²	2 Wires 31% in.²	Over 2 Wires 40% mm²	Over 2 Wires 40% in.²	1 Wire 53% mm²	1 Wire 53% in.²	60% mm²	60% in.²
12	3/8	—	—	—	—	—	—	—	—	—	—	—	—
16	1/2	15.3	0.602	184	0.285	57	0.088	74	0.114	97	0.151	110	0.171
21	3/4	20.4	0.804	327	0.508	101	0.157	131	0.203	173	0.269	196	0.305
27	1	26.1	1.029	535	0.832	166	0.258	214	0.333	284	0.441	321	0.499
35	1-1/4	34.5	1.360	935	1.453	290	0.450	374	0.581	495	0.770	561	0.872
41	1-1/2	40.4	1.590	1282	1.986	397	0.616	513	0.794	679	1.052	769	1.191
53	2	52.0	2.047	2124	3.291	658	1.020	849	1.316	1126	1.744	1274	1.975
63	2-1/2	62.1	2.445	3029	4.695	939	1.455	1212	1.878	1605	2.488	1817	2.817
78	3	77.3	3.042	4693	7.268	1455	2.253	1877	2.907	2487	3.852	2816	4.361
91	3-1/2	89.4	3.521	6277	9.737	1946	3.018	2511	3.895	3327	5.161	3766	5.842
103	4	101.5	3.998	8091	12.554	2508	3.892	3237	5.022	4288	6.654	4855	7.532

Continued

Exhibit VII.36 Dimensions and Percent Area of Rigid PVC Conduit (RNC), Schedule 40, and HDPE Conduit *Continued*

Metric Designator	Trade Size	Nominal Internal Diameter		Total Area 100%		2 Wires 31%		Over 2 Wires 40%		1 Wire 53%		60%	
		mm	in.	mm²	in.²	mm²	in.²	mm²	in.²	mm²	in.²	mm²	in.²
129	5	127.4	5.016	12748	19.761	3952	6.126	5099	7.904	6756	10.473	7649	11.856
155	6	153.2	6.031	18433	28.567	5714	8.856	7373	11.427	9770	15.141	11060	17.140

Note: Refer to the NEC for more information on conduit fill.
Source: NFPA 70, National Electrical Code®, NFPA, Quincy, MA, 2002 edition, Table 4.

Exhibit VII.37 Dimensions and Percent Area of Type A, Rigid PVC Conduit (RNC)

Metric Designator	Trade Size	Nominal Internal Diameter		Total Area 100%		2 Wires 31%		Over 2 Wires 40%		1 Wire 53%		60%	
		mm	in.	mm²	in.²	mm²	in.²	mm²	in.²	mm²	in.²	mm²	in.²
16	1/2	17.8	0.700	249	0.385	77	0.119	100	0.154	132	0.204	149	0.231
21	3/4	23.1	0.910	419	0.650	130	0.202	168	0.260	222	0.345	251	0.390
27	1	29.8	1.175	697	1.084	216	0.336	279	0.434	370	0.575	418	0.651
35	1-1/4	38.1	1.500	1140	1.767	353	0.548	456	0.707	604	0.937	684	1.060
41	1-1/2	43.7	1.720	1500	2.324	465	0.720	600	0.929	795	1.231	900	1.394
53	2	54.7	2.155	2350	3.647	728	1.131	940	1.459	1245	1.933	1410	2.188
63	2-1/2	66.9	2.635	3515	5.453	1090	1.690	1406	2.181	1863	2.890	2109	3.272
78	3	82.0	3.230	5281	8.194	1637	2.540	2112	3.278	2799	4.343	3169	4.916
91	3-1/2	93.7	3.690	6896	10.694	2138	3.315	2758	4.278	3655	5.668	4137	6.416
103	4	106.2	4.180	8858	13.723	2746	4.254	3543	5.489	4695	7.273	5315	8.234
129	5	—	—	—	—	—	—	—	—	—	—	—	—
155	6	—	—	—	—	—	—	—	—	—	—	—	—

Note: Refer to the *NEC* for more information on conduit fill.
Source: NFPA 70, *National Electrical Code®*, NFPA, Quincy, MA, 2002 edition, Table 4.

Exhibit VII.38 Dimensions and Percent Area of Type EB, PVC Conduit (RNC)

Metric Designator	Trade Size	Nominal Internal Diameter		Total Area 100%		2 Wires 31%		Over 2 Wires 40%		1 Wire 53%		60%	
		mm	in.	mm²	in.²	mm²	in.²	mm²	in.²	mm²	in.²	mm²	in.²
16	1/2	—	—	—	—	—	—	—	—	—	—	—	—
21	3/4	—	—	—	—	—	—	—	—	—	—	—	—
27	1	—	—	—	—	—	—	—	—	—	—	—	—
35	1-1/4	—	—	—	—	—	—	—	—	—	—	—	—
41	1-1/2	—	—	—	—	—	—	—	—	—	—	—	—
53	2	56.4	2.221	2498	3.874	774	1.201	999	1.550	1324	2.053	1499	2.325
63	2-1/2	—	—	—	—	—	—	—	—	—	—	—	—
78	3	84.6	3.330	5621	8.709	1743	2.700	2248	3.484	2979	4.616	3373	5.226
91	3-1/2	96.6	3.804	7329	11.365	2272	3.523	2932	4.546	3884	6.023	4397	6.819
103	4	108.9	4.289	9314	14.448	2887	4.479	3726	5.779	4937	7.657	5589	8.669
129	5	135.0	5.316	14314	22.195	4437	6.881	5726	8.878	7586	11.763	8588	13.317
155	6	160.9	6.336	20333	31.530	6303	9.774	8133	12.612	10776	16.711	12200	18.918

Note: Refer to the *NEC* for more information on conduit fill.
Source: NFPA 70, *National Electrical Code*®, NFPA, Quincy, MA; 2002 edition, Table 4.

Exhibit VII.39 Maximum Number of Same Size Conductors or Fixture Wires in Electrical Metallic Tubing (EMT)

	Conductor Size (AWG/kcmil)	Conductors									
		Metric Designator (Trade Size)									
Type		16 (1/2)	21 (3/4)	27 (1)	35 (1-1/4)	41 (1-1/2)	53 (2)	63 (2-1/2)	78 (3)	91 (3-1/2)	103 (4)
RHH, RHW, RHW-2	14	4	7	11	20	27	46	80	120	157	201
	12	3	6	9	17	23	38	66	100	131	167
	10	2	5	8	13	18	30	53	81	105	135
	8	1	2	4	7	9	16	28	42	55	70
	6	1	1	3	5	8	13	22	34	44	56
	4	1	1	2	4	6	10	17	26	34	44
	3	1	1	1	4	5	9	15	23	30	38
	2	1	1	1	3	4	7	13	20	26	33
	1	0	1	1	1	3	5	9	13	17	22
	1/0	0	1	1	1	2	4	7	11	15	19
	2/0	0	1	1	1	2	4	6	10	13	17
	3/0	0	0	1	1	1	3	5	8	11	14
	4/0	0	0	1	1	1	3	5	7	9	12

Continued

Exhibit VII.39 Maximum Number of Same Size Conductors or Fixture Wires in Electrical Metallic Tubing (EMT) *Continued*

Type	Conductor Size (AWG/kcmil)	Metric Designator (Trade Size)										
		16 (1/2)	21 (3/4)	27 (1)	35 (1-1/4)	41 (1-1/2)	53 (2)	63 (2-1/2)	78 (3)	91 (3-1/2)	103 (4)
RHH, RHW, RHW-2	250	0	0	0	1	1	1	3	5	7	9
	300	0	0	0	1	1	1	3	5	6	8
	350	0	0	0	1	1	1	2	4	6	7
	400	0	0	0	1	1	1	2	4	5	7
	500	0	0	0	0	1	1	2	3	4	6
	600	0	0	0	0	1	1	1	3	4	5
	700	0	0	0	0	0	1	1	2	3	4
	750	0	0	0	0	0	1	1	2	3	4
	800	0	0	0	0	0	1	1	2	3	4
	900	0	0	0	0	0	1	1	1	3	3
	1000	0	0	0	0	0	1	1	1	2	3
	1250	0	0	0	0	0	0	1	1	1	2
	1500	0	0	0	0	0	0	1	1	1	1
	1750	0	0	0	0	0	0	1	1	1	1
	2000	0	0	0	0	0	0	1	1	1	1

Type	Size										
TW, THHW, THW, THW-2	14	8	15	25	43	58	96	168	254	332	424
	12	6	11	19	33	45	74	129	195	255	326
	10	5	8	14	24	33	55	96	145	190	243
	8	2	5	8	13	18	30	53	81	105	135
RHH,[a] RHW,[a] RHW-2[a]	14	6	10	16	28	39	64	112	169	221	282
	12	4	8	13	23	31	51	90	136	177	227
	10	3	6	10	18	24	40	70	106	138	177
	8	1	4	6	10	14	24	42	63	83	106
RHH,[a] RHW,[a] RHW-2,[a] TW, THW, THHW, THW-2	6	1	3	4	8	11	18	32	48	63	81
	4	1	1	3	6	8	13	24	36	47	60
	3	1	1	3	5	7	12	20	31	40	52
	2	1	1	2	4	6	10	17	26	34	44
	1	1	1	1	3	4	7	12	18	24	31
	1/0	0	1	1	2	3	6	10	16	20	26
	2/0	0	1	1	1	2	5	9	13	17	22
	3/0	0	1	1	1	2	4	7	11	15	19
	4/0	0	0	1	1	1	3	6	9	12	16
	250	0	0	1	1	1	3	5	7	10	13
	300	0	0	1	1	1	2	4	6	8	11

Continued

Exhibit VII.39 Maximum Number of Same Size Conductors or Fixture Wires in Electrical Metallic Tubing (EMT) *Continued*

		Conductors									
		Metric Designator (Trade Size)									
Type	Conductor Size (AWG/kcmil)	16 (1/2)	21 (3/4)	27 (1)	35 (1-1/4)	41 (1-1/2)	53 (2)	63 (2-1/2)	78 (3)	91 (3-1/2)	103 (4)
RHH,[a] RHW,[a]	350	0	0	0	1	1	1	4	6	7	10
RHW-2,[a] TW,	400	0	0	0	1	1	1	3	5	7	9
THW, THHW,	500	0	0	0	1	1	1	3	4	6	7
THW-2	600	0	0	0	0	1	1	2	3	4	6
	700	0	0	0	0	1	1	1	3	4	5
	750	0	0	0	0	1	1	1	3	4	5
	800	0	0	0	0	1	1	1	3	3	5
	900	0	0	0	0	0	1	1	2	3	4
	1000	0	0	0	0	0	1	1	2	3	4
	1250	0	0	0	0	0	1	1	1	2	3
	1500	0	0	0	0	0	0	1	1	1	2
	1750	0	0	0	0	0	0	1	1	1	2
	2000	0	0	0	0	0	0	1	1	1	1

	12	22	35	61	84	138	241	364	476	608
THHN, THWN, THWN-2										
14	12	22	35	61	84	138	241	364	476	608
12	9	16	26	45	61	101	176	266	347	443
10	5	10	16	28	38	63	111	167	219	279
8	3	6	9	16	22	36	64	96	126	161
6	2	4	7	12	16	26	46	69	91	116
4	1	2	4	7	10	16	28	43	56	71
3	1	1	3	6	8	13	24	36	47	60
2	1	1	3	5	7	11	20	30	40	51
1	1	1	1	4	5	8	15	22	29	37
1/0	1	1	1	3	4	7	12	19	25	32
2/0	0	1	1	2	3	6	10	16	20	26
3/0	0	1	1	1	3	5	8	13	17	22
4/0	0	1	1	1	2	4	7	11	14	18
250	0	0	1	1	1	3	6	9	11	15
300	0	0	1	1	1	3	5	7	10	13
350	0	0	1	1	1	2	4	6	9	11
400	0	0	0	1	1	1	4	6	8	10
500	0	0	0	1	1	1	3	5	6	8
600	0	0	0	1	1	1	2	4	5	7
700	0	0	0	1	1	1	2	3	4	6

Continued

Exhibit VII.39 Maximum Number of Same Size Conductors or Fixture Wires in Electrical Metallic Tubing (EMT) *Continued*

Type	Conductor Size (AWG/kcmil)	Conductors Metric Designator (Trade Size)									
		16 (1/2)	21 (3/4)	27 (1)	35 (1-1/4)	41 (1-1/2)	53 (2)	63 (2-1/2)	78 (3)	91 (3-1/2)	103 (4)
THHN, THWN, THWN-2	750	0	0	0	0	1	1	1	3	4	5
	800	0	0	0	0	1	1	1	3	4	5
	900	0	0	0	0	1	1	1	3	3	4
	1000	0	0	0	0	1	1	1	2	3	4
FEP, FEPB, PF,a PFAH, TFE	14	12	21	34	60	81	134	234	354	462	590
	12	9	15	25	43	59	98	171	258	337	430
	10	6	11	18	31	42	70	122	185	241	309
	8	3	6	10	18	24	40	70	106	138	177
	6	2	4	7	12	17	28	50	75	98	126
	4	1	3	5	9	12	20	35	53	69	88
	3	1	2	4	7	10	16	29	44	57	73
	2	1	1	3	6	8	13	24	36	47	60
PFA, PFAH, TFE	1	1	1	2	4	6	9	16	25	33	42

Type	Size										
PFAH, TFE, PFA, PFAH, TFE, Z	1/0	1	1	1	3	5	8	14	21	27	35
	2/0	0	1	1	3	4	6	11	17	22	29
	3/0	0	1	1	2	3	5	9	14	18	24
	4/0	0	1	1	1	2	4	8	11	15	19
Z	14	14	25	41	72	98	161	282	426	556	711
	12	10	18	29	51	69	114	200	302	394	504
	10	6	11	18	31	42	70	122	185	241	309
	8	4	7	11	20	27	44	77	117	153	195
	6	3	5	8	14	19	31	54	82	107	137
	4	1	3	5	9	13	21	37	56	74	94
	3	1	2	4	7	9	15	27	41	54	69
	2	1	1	3	6	8	13	22	34	45	57
	1	1	1	2	4	6	10	18	28	36	46
XHH, XHHW, XHHW-2, ZW	14	8	15	25	43	58	96	168	254	332	424
	12	6	11	19	33	45	74	129	195	255	326
	10	5	8	14	24	33	55	96	145	190	243
	8	2	5	8	13	18	30	53	81	105	135
	6	1	3	6	10	14	22	39	60	78	100

Continued

Exhibit VII.39 Maximum Number of Same Size Conductors or Fixture Wires in Electrical Metallic Tubing (EMT) *Continued*

					Conductors					
					Metric Designator (Trade Size)					
Type	Conductor Size (AWG/kcmil)	16 (1/2)	21 (3/4)	27 (1)	35 (1-1/4)	41 (1-1/2)	53 (2)	63 (2-1/2)	78 (3)	91 (3-1/2)	103 (4)
XHH, XHHW, XHHW-2, ZW	4	1	2	4	7	10	16	28	43	56	72
	3	1	1	3	6	8	14	24	36	48	61
	2	1	1	3	5	7	11	20	31	40	51
XHH, XHHW, XHHW-2	1	1	1	1	4	5	8	15	23	30	38
	1/0	0	1	1	3	4	7	13	19	25	32
	2/0	0	1	1	2	3	6	10	16	21	27
	3/0	0	1	1	1	3	5	9	13	17	22
	4/0	0	1	1	1	2	4	7	11	14	18
	250	0	0	1	1	1	3	6	9	12	15
	300	0	0	1	1	1	3	5	8	10	13
	350	0	0	1	1	1	2	4	7	9	11
	400	0	0	0	1	1	1	4	6	8	10
	500	0	0	0	1	1	1	3	5	6	8

AWG/kcmil							27 (1)	35 (1-1/4)	41 (1-1/2)	53 (2)
600	0	0	0	0	0	1	1	2	4	5	6
700	0	0	0	0	0	0	1	2	3	4	6
750	0	0	0	0	0	0	1	1	3	4	5
800	0	0	0	0	0	0	1	1	3	4	5
900	0	0	0	0	0	0	1	1	3	3	4
1000	0	0	0	0	0	0	1	1	2	3	4
1250	0	0	0	0	0	0	1	1	1	2	3
1500	0	0	0	0	0	0	1	1	1	1	3
1750	0	0	0	0	0	0	0	1	1	1	2
2000	0	0	0	0	0	0	0	1	1	1	1

Fixture Wires

Type	Conductor Size (AWG/kcmil)	Metric Designator (Trade Size)					
		16 (1/2)	21 (3/4)	27 (1)	35 (1-1/4)	41 (1-1/2)	53 (2)
FFH-2, RFH-2, RFHH-3	18	8	14	24	41	56	92
	16	7	12	20	34	47	78
SF-2, SFF-2	18	10	18	30	52	71	116
	16	8	15	25	43	58	96

Continued

Exhibit VII.39 Maximum Number of Same Size Conductors or Fixture Wires in Electrical Metallic Tubing (EMT) *Continued*

Fixture Wires

Type	Conductor Size (AWG/kcmil)	Metric Designator (Trade Size)					
		16 (1/2)	21 (3/4)	27 (1)	35 (1-1/4)	41 (1-1/2)	53 (2)
SF-1, SFF-1	14	7	12	20	34	47	78
	18	18	33	53	92	125	206
RFH-1, RFHH-2, TF, TFF, XF, XFF	18	14	24	39	68	92	152
RFHH-2, TF, TFF, XF, XFF	16	11	19	31	55	74	123
XF, XFF	14	8	15	25	43	58	96
TFN, TFFN	18	22	38	63	108	148	244
	16	17	29	48	83	113	186
PF, PFF, PGF, PGFF, PAF, PTF, PTFF, PAFF	18	21	36	59	103	140	231
	16	16	28	46	79	108	179
	14	12	21	34	60	81	134
ZF, ZFF, ZHF, HF, HFF	18	27	47	77	113	181	298
	16	20	35	56	83	133	220
	14	14	25	51	72	98	161

Type	Size						
KF-2, KFF-2	18	39	69	111	193	262	433
	16	27	48	78	136	185	305
	14	19	33	54	93	127	209
	12	13	23	37	64	87	144
	10	8	15	25	43	58	96
KF-1, KFF-1	18	46	82	133	230	313	516
	16	33	57	93	161	220	362
	14	22	38	63	108	148	244
	12	14	25	41	72	98	161
	10	9	16	27	47	64	105
XF, XFF	12	4	8	13	23	31	51
	10	3	6	10	18	24	40

Note: This table is for concentric stranded conductors only. For compact stranded conductors, Exhibit VII.40 should be used.

[a]Types RHH, RHW, and RHW-2 without outer covering.

Source: NFPA 70, *National Electrical Code*®, NFPA, Quincy, MA, 2002 edition, Table C1.

Exhibit VII.40 Maximum Number of Same Size Compact Conductors in Electrical Metallic Tubing (EMT)

Compact Conductors

Type	Conductor Size (AWG/kcmil)	Metric Designator (Trade Size)									
		16 (1/2)	21 (3/4)	27 (1)	35 (1-1/4)	41 (1-1/2)	53 (2)	63 (2-1/2)	78 (3)	91 (3-1/2)	103 (4)
THW, THW-2, THHW	8	2	4	6	11	16	26	46	69	90	115
	6	1	3	5	9	12	20	35	53	70	89
	4	1	2	4	6	9	15	26	40	52	67
	2	1	1	3	5	7	11	19	29	38	49
	1	1	1	3	3	4	8	13	21	27	34
	1/0	1	1	1	3	4	7	12	18	23	30
	2/0	0	1	1	2	3	5	10	15	20	25
	3/0	0	1	1	1	3	5	8	13	17	21
	4/0	0	1	1	1	2	4	7	11	14	18
	250	0	0	1	1	1	3	5	8	11	14
	300	0	0	1	1	1	3	5	7	9	12
	350	0	0	1	1	1	2	4	6	8	11
	400	0	0	0	1	1	2	4	6	8	10
	500	0	0	0	1	1	1	3	5	6	8

Size										
600	0	0	0	0	0	1	1	1	1	2	4	7
700	0	0	0	0	1	1	1	2	2	3	4	6
750	0	0	0	0	0	1	1	1	1	3	4	5
1000	0	0	0	0	0	1	1	1	1	2	3	4

THHN, THWN, THWN-2

Size										
8	—	—	—	—	—	—	—	—	—	—
6	2	4	7	13	18	29	52	78	102	130
4	1	3	4	8	11	18	32	48	63	81
2	1	1	3	6	8	13	23	34	45	58
1	1	1	2	4	6	10	17	26	34	43
1/0	1	1	1	3	5	8	14	22	29	37
2/0	1	1	1	3	4	7	12	18	24	30
3/0	0	1	1	2	3	6	10	15	20	25
4/0	0	1	1	1	3	5	8	12	16	21
250	0	1	1	1	1	4	6	10	13	16
300	0	0	1	1	1	3	5	8	11	14
350	0	0	1	1	1	3	5	7	10	12
400	0	0	1	1	1	2	4	6	9	11

Continued

Exhibit VII.40 Maximum Number of Same Size Compact Conductors in Electrical Metallic Tubing (EMT) Continued

Compact Conductors

Type	Conductor Size (AWG/kcmil)	16 (1/2)	21 (3/4)	27 (1)	35 (1-1/4)	41 (1-1/2)	53 (2)	63 (2-1/2)	78 (3)	91 (3-1/2)	103 (4)
							Metric Designator (Trade Size)				
THHN, THWN, THWN-2	500	0	0	0	1	1	1	4	5	7	9
	600	0	0	0	1	1	1	3	4	6	7
	700	0	0	0	1	1	1	2	4	5	7
	750	0	0	0	1	1	1	2	4	5	6
	1000	0	0	0	0	1	1	1	3	3	4
XHHW, XHHW-2	8	3	5	8	15	20	34	59	90	117	149
	6	1	4	6	11	15	25	44	66	87	111
	4	1	3	4	8	11	18	32	48	63	81
	2	1	1	3	6	8	13	23	34	45	58
	1	1	1	2	4	6	10	17	26	34	43
	1/0	1	1	1	3	5	8	14	22	29	37
	2/0	1	1	1	3	4	7	12	18	24	31
	3/0	0	1	1	2	3	6	10	15	20	25
	4/0	0	0	1	1	3	5	8	13	13	21

250	0	1	1	2	4	7	10	13	17
300	0	0	1	1	3	6	9	11	14
350	0	0	1	1	3	5	8	10	13
400	0	0	1	1	2	4	7	9	11
500	0	0	0	1	1	4	6	7	9
600	0	0	0	1	1	3	4	6	8
700	0	0	0	1	1	2	4	5	7
750	0	0	0	1	1	2	3	5	6
1000	0	0	0	0	1	1	3	4	5

Note: *Compact stranding* is defined as the result of a manufacturing process where the standard conductor is compressed to the extent that the interstices (voids between strand wires) are virtually eliminated.

Source: NFPA 70, *National Electrical Code®*, NFPA, Quincy, MA, 2002 edition, Table C1(A).

CONDUIT AND RACEWAY SYSTEMS

Exhibit VII.41 Maximum Number of Same Size Conductors or Fixture Wires in Electrical Nonmetallic Tubing (ENT)

Type	Conductor Size (AWG/kcmil)	Conductors — Metric Designator (Trade Size)					
		16 (1/2)	21 (3/4)	27 (1)	35 (1-1/4)	41 (1-1/2)	53 (2)
RHH, RHW, RHW-2	14	3	6	10	19	26	43
	12	2	5	9	16	22	36
RHH, RHW, RHW-2	10	1	4	7	13	17	29
	8	1	1	3	6	9	15
	6	1	1	3	5	7	12
	4	1	1	2	4	6	9
	3	1	1	1	3	5	8
	2	0	1	1	3	4	7
	1	0	1	1	1	3	5
	1/0	0	0	1	1	2	4
	2/0	0	0	1	1	1	3
	3/0	0	0	1	1	1	3
	4/0	0	0	1	1	1	2
	250	0	0	0	1	1	1
	300	0	0	0	1	1	1

Size						
350	1	1	1	0	0	0
400	1	1	1	0	0	0
500	1	1	0	0	0	0
600	1	1	0	0	0	0
700	1	0	0	0	0	0
750	1	0	0	0	0	0
800	1	0	0	0	0	0
900	1	0	0	0	0	0
1000	1	0	0	0	0	0
1250	0	0	0	0	0	0
1500	0	0	0	0	0	0
1750	0	0	0	0	0	0
2000	0	0	0	0	0	0
TW, THHW, THW, THW-2						
14	92	55	40	22	13	7
12	71	42	31	17	10	5
10	52	32	23	13	7	4
8	29	17	13	7	4	1
RHH,[a] RHW,[a] RHW-2[a]						
14	61	37	27	15	8	4

Continued

Exhibit VII.41 Maximum Number of Same Size Conductors or Fixture Wires in Electrical Nonmetallic Tubing (ENT) *Continued*

Type	Conductor Size (AWG/kcmil)	Conductors Metric Designator (Trade Size)					
		16 (1/2)	21 (3/4)	27 (1)	35 (1-1/4)	41 (1-1/2)	53 (2)
RHH,[a] RHW,[a] RHW-2[a]	12	3	7	12	21	29	49
	10	3	5	9	17	23	38
RHH,[a] RHW,[a] RHW-2[a]	8	1	3	5	10	14	23
RHH,[a] RHW,[a] RHW-2,[a] TW, THW, THHW, THW-2	6	1	2	4	7	10	17
	4	1	1	3	5	8	13
	3	1	1	3	5	7	11
	2	1	1	2	4	6	9
	1	0	1	1	3	4	6
	1/0	0	1	1	2	3	5
	2/0	0	1	1	1	3	5
	3/0	0	0	1	1	2	4
	4/0	0	0	1	1	1	3

250	0	0	1	1	1	2
300	0	0	0	1	1	2
350	0	0	0	1	1	1
400	0	0	0	1	1	1
500	0	0	0	1	1	1
600	0	0	0	0	1	1
700	0	0	0	0	1	1
750	0	0	0	0	1	1
800	0	0	0	0	1	1
900	0	0	0	0	0	1
1000	0	0	0	0	0	1
1250	0	0	0	0	0	1
1500	0	0	0	0	0	0
1750	0	0	0	0	0	0
2000	0	0	0	0	0	0
14	10	18	32	58	80	132
12	7	13	23	42	58	96
10	4	8	15	26	36	60
8	2	5	8	15	21	35
6	1	3	6	11	15	25

THHN, THWN, THWN-2

Continued

Exhibit VII.41 Maximum Number of Same Size Conductors or Fixture Wires in Electrical Nonmetallic Tubing (ENT) *Continued*

Type	Conductor Size (AWG/kcmil)	Conductors					
		Metric Designator (Trade Size)					
		16 (1/2)	21 (3/4)	27 (1)	35 (1-1/4)	41 (1-1/2)	53 (2)
THHN, THWN, THWN-2	4	1	1	4	7	9	15
	3	1	1	3	5	8	13
	2	1	1	2	5	6	11
	1	1	1	1	3	5	8
	1/0	0	1	1	3	4	7
	2/0	0	1	1	2	3	5
	3/0	0	1	1	1	3	4
	4/0	0	0	1	1	2	4
	250	0	0	1	1	1	3
	300	0	0	1	1	1	2
	350	0	0	0	1	1	2
	400	0	0	0	1	1	1
	500	0	0	0	1	1	1

Type	Size						
	600	0	0	0	1	1	1
	700	0	0	0	0	1	1
	750	0	0	0	0	1	1
	800	0	0	0	0	1	1
	900	0	0	0	0	1	1
	1000	0	0	0	0	0	1
FEP, FEPB, PFA, PFAH, TFE	14	10	18	31	56	77	128
	12	7	13	23	41	56	93
	10	5	9	16	29	40	67
	8	3	5	9	17	23	38
	6	1	4	6	12	16	27
	4	1	2	4	8	11	19
	3	1	1	4	7	9	16
	2	1	1	3	5	8	13
PFA, PFAH, TFE	1	1	1	1	4	5	9
PFA, PFAH, TFE, Z	1/0	0	1	1	3	4	7
	2/0	0	1	1	2	4	6
	3/0	0	1	1	1	3	5
	4/0	0	1	1	1	2	4

Continued

Exhibit VII.41 Maximum Number of Same Size Conductors or Fixture Wires in Electrical Nonmetallic Tubing (ENT) *Continued*

Type	Conductor Size (AWG/kcmil)	Conductors Metric Designator (Trade Size)					
		16 (1/2)	21 (3/4)	27 (1)	35 (1-1/4)	41 (1-1/2)	53 (2)
Z	14	12	22	38	68	93	154
	12	8	15	27	48	66	109
	10	5	9	16	29	40	67
	8	3	6	10	18	25	42
	6	1	4	7	13	18	30
	4	1	3	5	9	12	20
	3	1	1	3	6	9	15
	2	1	1	3	5	7	12
	1	1	1	2	4	6	10
XHH, XHHW, XHHW-2, ZW	14	7	13	22	40	55	92
	12	5	10	17	31	42	71
	10	4	7	13	23	32	52
	8	1	4	7	13	17	29
	6	1	3	5	9	13	21

XHH, XHHW,
XHHW-2

Size						
4	1	1	4	7	9	15
3	1	1	3	6	8	13
2	1	1	2	5	6	11
1	1	1	1	3	5	8
1/0	0	1	1	3	4	7
2/0	0	1	1	2	3	6
3/0	0	1	1	1	3	5
4/0	0	0	1	1	2	4
250	0	0	1	1	1	3
300	0	0	1	1	1	3
350	0	0	0	1	1	2
400	0	0	0	1	1	1
500	0	0	0	1	1	1
600	0	0	0	1	1	1
700	0	0	0	0	1	1
750	0	0	0	0	1	1
800	0	0	0	0	1	1
900	0	0	0	0	1	1

Continued

Exhibit VII.41 Maximum Number of Same Size Conductors or Fixture Wires in Electrical Nonmetallic Tubing (ENT) *Continued*

Conductors

Type	Conductor Size (AWG/kcmil)	Metric Designator (Trade Size)					
		16 (1/2)	21 (3/4)	27 (1)	35 (1-1/4)	41 (1-1/2)	53 (2)
XHH, XHHW, XHHW-2	1000	0	0	0	0	0	1
	1250	0	0	0	0	0	1
	1500	0	0	0	0	0	1
	1750	0	0	0	0	0	0
	2000	0	0	0	0	0	0

Fixture Wires

Type	Conductor Size (AWG/kcmil)	Metric Designator (Trade Size)					
		16 (1/2)	21 (3/4)	27 (1)	35 (1-1/4)	41 (1-1/2)	53 (2)
FFH-2, RFH-2, RFHH-3	18	6	12	21	39	53	88
	16	5	10	18	32	45	74

Type	Size						
SF-2, SFF-2	18	8	15	27	49	67	111
	16	7	13	22	40	55	92
	14	5	10	18	32	45	74
SF-1, SFF-1	18	15	28	48	86	119	197
RFH-1, RFHH-2, TF, TFF, XF, XFF	18	11	20	35	64	88	145
RFHH-2, TF, TFF, XF, XFF	16	9	16	29	51	71	117
XF, XFF	14	7	13	22	40	55	92
TFN, TFFN	18	18	33	57	102	141	233
	16	13	25	43	78	107	178
PF, PFF, PGF, PGFF, PAF, PTF, PTFF, PAFF	18	17	31	54	97	133	221
	16	13	24	42	75	103	171
	14	10	18	31	56	77	128
ZF, ZFF, ZHF, HF, HFF	18	22	40	70	125	172	285
	16	16	29	51	92	127	210
	14	12	22	38	68	93	154
KF-2, KFF-2	18	31	58	101	182	250	413
	16	22	41	71	128	176	291

Continued

Exhibit VII.41 Maximum Number of Same Size Conductors or Fixture Wires in Electrical Nonmetallic Tubing (ENT) *Continued*

Fixture Wires

Type	Conductor Size (AWG/kcmil)	Metric Designator (Trade Size)					
		16 (1/2)	21 (3/4)	27 (1)	35 (1-1/4)	41 (1-1/2)	53 (2)
KF-2, KFF-2	14	15	28	49	88	121	200
	12	10	19	33	60	83	138
	10	7	13	22	40	55	92
KF-1, KFF-1	18	38	69	121	217	298	493
	16	26	49	85	152	209	346
	14	18	33	57	102	141	233
	12	12	22	38	68	93	154
	10	7	14	24	44	61	101
XF, XFF	12	3	7	12	21	29	49
	10	3	5	9	17	23	38

Note: This table is for concentric stranded conductors only. For compact stranded conductors, Exhibit VII.42 should be used.

[a]Types RHH, RHW, and RHW-2 without outer covering.

Source: NFPA 70, *National Electrical Code®*, NFPA, Quincy, MA, 2002 edition, Table C2.

Exhibit VII.42 Maximum Number of Same Size Compact Conductors in Electrical Nonmetallic Tubing (ENT)

Type	Conductor Size (AWG/kcmil)	Metric Designator (Trade Size)						
		16 (1/2)	21 (3/4)	27 (1)	35 (1-1/4)	41 (1-1/2)	53 (2)
THW, THW-2, THHW	8	1	3	6	11	15	25
	6	1	2	4	8	11	19
	4	1	1	3	6	8	14
	2	1	1	2	4	6	10
	1	0	1	1	3	4	7
	1/0	0	1	1	3	4	6
	2/0	0	1	1	2	3	5
	3/0	0	1	1	1	3	4
	4/0	0	0	1	1	2	4
	250	0	0	1	1	1	3
	300	0	0	1	1	1	2
	350	0	0	0	1	1	2
	400	0	0	0	1	1	1
	500	0	0	0	1	1	1

Continued

Exhibit VII.42 Maximum Number of Same Size Compact Conductors in Electrical Nonmetallic Tubing (ENT) *Continued*

	Compact Conductors							
	Conductor Size	Metric Designator (Trade Size)						
Type	(AWG/kcmil)	16 (1/2)	21 (3/4)	27 (1)	35 (1-1/4)	41 (1-1/2)	53 (2)	
THW, THW-2, THHW	600	0	0	0	1	1	1	
	700	0	0	0	1	1	1	
	750	0	0	0	1	1	1	
	1000	0	0	0	0	0	1	
THHN, THWN, THWN-2	8	—	—	—	—	—	28	
	6	1	4	7	12	17	17	
	4	1	2	4	7	10	12	
	2	1	1	3	5	7	9	
	1	1	1	2	4	5	9	
	1/0	1	1	1	3	5	8	
	2/0	0	1	1	3	4	6	
	3/0	0	1	1	2	3	5	
	4/0	0	1	1	1	2	4	

Continued

Type	Size						
XHHW, XHHW-2	250	3	1	1	1	0	0
	300	3	1	1	1	0	0
	350	2	1	1	0	0	0
	400	2	1	1	0	0	0
	500	1	1	1	0	0	0
	600	1	1	1	0	0	0
	700	1	1	1	0	0	0
	750	1	1	1	0	0	0
	1000	1	1	0	0	0	0
	8	32	19	14	8	4	2
	6	24	14	10	6	3	1
	4	17	10	7	4	2	1
	2	12	7	5	3	1	1
	1	9	5	4	2	1	1
	1/0	8	5	3	1	1	1
	2/0	7	4	3	1	1	0
	3/0	5	3	2	1	1	0
	4/0	4	3	1	1	1	0

Exhibit VII.42 Maximum Number of Same Size Compact Conductors in Electrical Nonmetallic Tubing (ENT) – Continued

	Compact Conductors	Metric Designator (Trade Size)						
Type	Conductor Size (AWG/kcmil)	16 (1/2)	21 (3/4)	27 (1)	35 (1-1/4)	41 (1-1/2)	53 (2)
XHHW, XHHW-2	250	0	0	1	1	1	3
	300	0	0	1	1	1	3
	350	0	0	1	1	1	3
	400	0	0	1	1	1	2
	500	0	0	0	1	1	1
	600	0	0	0	1	1	1
	700	0	0	0	1	1	1
	750	0	0	0	1	1	1
	1000	0	0	0	0	1	1

Note: *Compact stranding* is defined as the result of a manufacturing process where the standard conductor is compressed to the extent that the interstices (voids between strand wires) are virtually eliminated.

Source: NFPA 70, *National Electrical Code®*, NFPA, Quincy, MA, 2002 edition, Table C2(A).

Exhibit VII.43 Maximum Number of Same Size Conductors or Fixture Wires in Flexible Metal Conduit (FMC)

Type	Conductor Size (AWG/kcmil)	Metric Designator (Trade Size)										
		16 (1/2)	21 (3/4)	27 (1)	35 (1-1/4)	41 (1-1/2)	53 (2)	63 (2-1/2)	78 (3)	91 (3-1/2)	103 (4)
RHH, RHW, RHW-2	14	4	7	11	17	25	44	67	96	131	171
	12	3	6	9	14	21	37	55	80	109	142
RHH, RHW, RHW-2	10	3	5	7	11	17	30	45	64	88	115
	8	1	2	4	6	9	15	23	34	46	60
	6	1	1	3	5	7	12	19	27	37	48
	4	1	1	2	4	5	10	14	21	29	37
	3	1	1	1	3	5	8	13	18	25	33
	2	1	1	1	3	4	7	11	16	22	28
	1	0	1	1	1	2	5	7	10	14	19
	1/0	0	1	1	1	2	4	6	9	12	16
	2/0	0	1	1	1	1	3	5	8	11	14

Continued

Exhibit VII.43 Maximum Number of Same Size Conductors or Fixture Wires in Flexible Metal Conduit (FMC) *Continued*

Type	Conductor Size (AWG/kcmil)	Conductors — Metric Designator (Trade Size)									
		16 (1/2)	21 (3/4)	27 (1)	35 (1-1/4)	41 (1-1/2)	53 (2)	63 (2-1/2)	78 (3)	91 (3-1/2)	103 (4)
RHH, RHW, RHW-2	3/0	0	0	1	1	1	3	5	7	9	12
	4/0	0	0	1	1	1	2	4	6	8	10
	250	0	0	0	1	1	1	3	4	6	8
	300	0	0	0	1	1	1	2	4	5	7
	350	0	0	0	1	1	1	2	3	5	6
	400	0	0	0	0	0	1	1	3	4	6
	500	0	0	0	0	0	1	1	3	4	5
	600	0	0	0	0	1	1	1	2	3	4
	700	0	0	0	0	0	1	1	1	3	3
	750	0	0	0	0	0	1	1	1	2	3
	800	0	0	0	0	0	1	1	1	2	3
	900	0	0	0	0	0	1	1	1	2	3

Type	Size										
	1000	0	0	0	0	0	1	1	1	1	3
	1250	0	0	0	0	0	0	1	1	1	1
	1500	0	0	0	0	0	0	1	1	1	1
	1750	0	0	0	0	0	0	1	1	1	1
	2000	0	0	0	0	0	0	0	1	1	1
TW, THHW, THW, THW-2	14	9	15	23	36	53	94	141	203	277	361
	12	7	11	18	28	41	72	108	156	212	277
	10	5	8	13	21	30	54	81	116	158	207
	8	3	5	7	11	17	30	45	64	88	115
RHH,[a] RHW,[a] RHW-2[a]	14	6	10	15	24	35	62	94	135	184	240
RHH,[a] RHW,[a] RHW-2[a]	12	5	8	12	19	28	50	75	108	148	193
	10	4	6	10	15	22	39	59	85	115	151
RHH,[a] RHW,[a] RHW-2[a]	8	1	4	6	9	13	23	35	51	69	90

Continued

Exhibit VII.43 Maximum Number of Same Size Conductors or Fixture Wires in Flexible Metal Conduit (FMC) *Continued*

Type	Conductor Size (AWG/kcmil)	Metric Designator (Trade Size)									
		16 (1/2)	21 (3/4)	27 (1)	35 (1-1/4)	41 (1-1/2)	53 (2)	63 (2-1/2)	78 (3)	91 (3-1/2)	103 (4)
RHH,[a] RHW,[a] RHW-2,[a] TW, THW, THHW, THW-2	6	1	3	4	7	10	18	27	39	53	69
	4	1	1	3	5	7	13	20	29	39	51
	3	1	1	3	4	6	11	17	25	34	44
	2	1	1	2	4	5	10	14	21	29	37
	1	1	1	1	2	4	7	10	15	20	26
	1/0	0	1	1	1	3	6	9	12	17	22
	2/0	0	1	1	1	3	5	7	10	14	19
	3/0	0	0	1	1	2	4	6	9	12	16
	4/0	0	0	1	1	1	3	5	7	10	13
	250	0	0	1	1	1	2	4	6	8	11
	300	0	0	1	1	1	2	3	5	7	9
	350	0	0	0	1	1	1	3	4	6	8

Size										
400	0	0	0	1	1	1	3	4	6	7
500	0	0	0	1	1	1	2	3	5	6
600	0	0	0	0	1	1	1	3	4	5
700	0	0	0	0	1	1	1	2	3	4
750	0	0	0	0	1	1	1	2	3	4
800	0	0	0	0	1	1	1	1	3	4
900	0	0	0	0	0	1	1	1	3	3
1000	0	0	0	0	0	1	1	1	2	3
1250	0	0	0	0	0	0	1	1	1	2
1500	0	0	0	0	0	0	1	1	1	1
1750	0	0	0	0	0	0	1	1	1	1
2000	0	0	0	0	0	0	1	1	1	1
14	13	22	33	52	76	134	202	291	396	518
12	9	16	24	38	56	98	147	212	289	378
10	6	10	15	24	35	62	93	134	182	238
8	3	6	9	14	20	35	53	77	105	137
6	2	4	6	10	14	25	38	55	76	99
4	1	2	4	6	9	16	24	34	46	61
3	1	1	3	5	7	13	20	29	39	51

THHN, THWN, THWN-2

Continued

Exhibit VII.43 Maximum Number of Same Size Conductors or Fixture Wires in Flexible Metal Conduit (FMC) *Continued*

	Conductor Size (AWG/kcmil)	Conductors									
		Metric Designator (Trade Size)									
Type		16 (1/2)	21 (3/4)	27 (1)	35 (1-1/4)	41 (1-1/2)	53 (2)	63 (2-1/2)	78 (3)	91 (3-1/2)	103 (4)
THHN, THWN, THWN-2	2	1	1	3	4	6	11	17	24	33	43
	1	1	1	1	3	4	8	12	18	24	32
	1/0	1	1	1	2	4	7	10	15	20	27
	2/0	0	1	1	1	3	6	9	12	17	22
	3/0	0	1	1	1	2	5	7	10	14	18
	4/0	0	1	1	1	1	4	6	8	12	15
	250	0	0	1	1	1	3	5	7	9	12
	300	0	0	1	1	1	3	4	6	8	11
	350	0	0	1	1	1	2	3	5	7	9
	400	0	0	0	1	1	1	3	5	6	8
	500	0	0	0	1	1	1	2	4	5	7
	600	0	0	0	0	1	1	1	3	4	5
	700	0	0	0	0	1	1	1	3	4	5

Type	Conductor Size										
	750	0	0	0	0	1	1	1	2	3	4
	800	0	0	0	0	1	1	1	2	3	4
	900	0	0	0	0	0	1	1	1	3	4
	1000	0	0	0	0	0	1	1	1	3	3
FEP, FEPB, PFA, PFAH, TFE	14	12	21	32	51	74	130	196	282	385	502
	12	9	15	24	37	54	95	143	206	281	367
	10	6	11	17	26	39	68	103	148	201	263
	8	4	6	10	15	22	39	59	85	115	151
	6	2	4	7	11	16	28	42	60	82	107
	4	1	3	5	7	11	19	29	42	57	75
	3	1	2	4	6	9	16	24	35	48	62
	2	1	1	3	5	7	13	20	29	39	51
PFA, PFAH, TFE	1	1	1	2	3	5	9	14	20	27	36
PFA, PFAH, TFE, Z	1/0	1	1	1	3	4	8	11	17	23	30
	2/0	1	1	1	2	3	6	9	14	19	24
	3/0	0	1	1	1	3	5	8	11	15	20
	4/0	0	1	1	1	2	4	6	9	13	16

Continued

Exhibit VII.43 Maximum Number of Same Size Conductors or Fixture Wires in Flexible Metal Conduit (FMC) *Continued*

Conductors

Type	Conductor Size (AWG/kcmil)	Metric Designator (Trade Size)									
		16 (1/2)	21 (3/4)	27 (1)	35 (1-1/4)	41 (1-1/2)	53 (2)	63 (2-1/2)	78 (3)	91 (3-1/2)	103 (4)
Z	14	15	25	39	61	89	157	236	340	463	605
	12	11	18	28	43	63	111	168	241	329	429
	10	6	11	17	26	39	68	103	148	201	263
	8	4	7	11	17	24	43	65	93	127	166
	6	3	5	7	12	17	30	45	65	89	117
	4	1	3	5	8	12	21	31	45	61	80
	3	1	2	4	6	8	15	23	33	45	58
	2	1	1	3	5	7	12	19	27	37	49
	1	1	1	2	4	6	10	15	22	30	39
XHH, XHHW, XHHW-2, ZW	14	9	15	23	36	53	94	141	203	277	361
	12	7	11	18	28	41	72	108	156	212	277
	10	5	8	13	21	30	54	81	116	158	207

Continued

XHH, XHHW, XHHW-2

Size										
8	115	88	64	45	30	17	11	7	5	3
6	85	65	48	33	22	12	8	5	3	1
4	61	47	34	24	16	9	6	4	2	1
3	52	40	29	20	13	7	5	3	1	1
2	44	33	24	17	11	6	4	3	1	1
1	32	25	18	13	8	5	3	1	1	1
1/0	27	21	15	10	7	4	2	1	1	1
2/0	23	17	13	9	6	3	2	1	1	0
3/0	19	14	10	7	5	3	1	1	1	0
4/0	15	12	9	6	4	2	1	1	1	0
250	13	10	7	5	3	1	1	1	0	0
300	11	8	6	4	3	1	1	1	0	0
350	9	7	5	4	2	1		0	0	0
400	8	6	5	3	1	1	1	0	0	0
500	7	5	4	3	1	1		0	0	0
600	5	4	3	1	1	1	0	0	0	0
700	5	4	3	1	1	1	0	0	0	0

Exhibit VII.43 Maximum Number of Same Size Conductors or Fixture Wires in Flexible Metal Conduit (FMC) *Continued*

	Conductor Size (AWG/kcmil)	Conductors									
		Metric Designator (Trade Size)									
Type		16 (1/2)	21 (3/4)	27 (1)	35 (1-1/4)	41 (1-1/2)	53 (2)	63 (2-1/2)	78 (3)	91 (3-1/2)	103 (4)
XHH, XHHW, XHHW-2	750	0	0	0	0	1	1	1	2	3	4
	800	0	0	0	0	1	1	1	2	3	4
	900	0	0	0	0	0	1	1	1	3	4
	1000	0	0	0	0	0	1	1	1	3	3
	1250	0	0	0	0	0	0	1	1	1	3
	1500	0	0	0	0	0	0	1	1	1	2
	1750	0	0	0	0	0	0	1	1	1	1
	2000	0	0	0	0	0	0	0	1	1	1

aTypes RHH, RHW, and RHW-2 without outer covering.

Fixture Wires

Type	Conductor Size (AWG/kcmil)	Metric Designator (Trade Size)					
		16 (1/2)	21 (3/4)	27 (1)	35 (1-1/4)	41 (1-1/2)	53 (2)
FFH-2, RFH-2, RFHH-3	18	8	14	22	35	51	90
	16	7	12	19	29	43	76
SF-2, SFF-2	18	11	18	28	44	64	113
	16	9	15	23	36	53	94
	14	7	12	19	29	43	76
SF-1, SFF-1	18	19	32	50	78	114	201
RFH-1, RFHH-2, TF, TFF, XF, XFF	18	14	24	37	58	84	148
RFHH-2, TF, TFF, XF, XFF	16	11	19	30	47	68	120
XF, XFF	14	9	15	23	36	53	94

Continued

Exhibit VII.43 Maximum Number of Same Size Conductors or Fixture Wires in Flexible Metal Conduit (FMC) *Continued*

Fixture Wires

Type	Conductor Size (AWG/kcmil)	Metric Designator (Trade Size)					
		16 (1/2)	21 (3/4)	27 (1)	35 (1-1/4)	41 (1-1/2)	53 (2)
TFN, TFFN	18	23	38	59	93	135	237
	16	17	29	45	71	103	181
PF, PFF, PGF, PGFF, PAF, PTF, PTFF, PAFF	18	22	36	56	88	128	225
	16	17	28	43	68	99	174
	14	12	21	32	51	74	130
ZF, ZFF, ZHF, HF, HFF	18	28	47	72	113	165	290
	16	20	35	53	83	121	214
	14	15	25	39	61	89	157
KF-2, KFF-2	18	41	68	105	164	239	421
	16	28	48	74	116	168	297
	14	19	33	51	80	116	204
	12	13	23	35	55	80	140
	10	9	15	23	36	53	94

Type	Size						
KF-1, KFF-1	18	48	82	125	196	285	503
	16	34	57	88	138	200	353
	14	23	38	59	93	135	237
	12	15	25	39	61	89	157
	10	10	16	25	40	58	103
XF, XFF	12	5	8	12	19	28	50
	10	4	6	10	15	22	39

Note: This table is for concentric stranded conductors only. For compact stranded conductors, Exhibit VII.44 should be used.

Source: NFPA 70, *National Electrical Code*®, NFPA, Quincy, MA, 2002 edition, Table C3.

Exhibit VII.44 Maximum Number of Same Size Compact Conductors in Flexible Metal Conduit (FMC)

Compact Conductors

Type	Conductor Size (AWG/kcmil)	Metric Designator (Trade Size)									
		16 (1/2)	21 (3/4)	27 (1)	35 (1-1/4)	41 (1-1/2)	53 (2)	63 (2-1/2)	78 (3)	91 (3-1/2)	103 (4)
THW, THHW, THW-2	8	2	4	6	10	14	25	38	55	75	98
	6	1	3	5	7	11	20	29	43	58	76
	4	1	2	3	5	8	15	22	32	43	57
	2	1	1	2	4	6	11	16	23	32	42
	1	1	1	1	3	4	7	11	16	22	29
	1/0	1	1	1	2	3	6	10	14	19	25
	2/0	0	1	1	1	3	5	8	12	16	21
	3/0	0	1	1	1	2	4	7	10	14	18
	4/0	0	1	1	1	1	4	6	8	11	15
	250	0	0	1	1	1	3	4	7	9	12
	300	0	0	1	1	1	2	4	6	8	10
	350	0	0	1	1	1	2	3	5	7	9
	400	0	0	0	1	1	2	3	5	6	8
	500	0	0	0	1	1	1	3	4	5	7

THHN, THWN, THWN-2

600	6	4	3	1	1	1	0	0	0	0
700	5	4	3	1	1	1	0	0	0	0
750	5	3	2	1	1	1	0	0	0	0
1000	4	3	1	1	1	0	0	0	0	0
8	—	—	—	—	—	—	—	—	—	—
6	111	85	62	43	29	16	11	7	4	3
4	69	52	38	27	18	10	7	4	3	1
2	49	38	28	19	13	7	5	3	1	1
1	37	28	21	14	9	5	3	2	1	1
1/0	31	24	17	12	8	4	3	1	1	1
2/0	26	20	14	10	6	4	2	1	1	1
3/0	22	17	12	8	5	3	1	1	1	0
4/0	18	14	10	7	4	2	1	1	1	0
250	14	11	8	5	3	1	1	1	1	0
300	12	9	7	5	3	1	1	1	0	0
350	10	8	6	4	3	1	1	1	0	0
400	9	7	5	3	2	1	1	1	0	0
500	8	6	4	3	1	1	1	0	0	0

Continued

Exhibit VII.44 Maximum Number of Same Size Compact Conductors in Flexible Metal Conduit (FMC) *Continued*

Type	Conductor Size (AWG/kcmil)	Compact Conductors Metric Designator (Trade Size)									
		16 (1/2)	21 (3/4)	27 (1)	35 (1-1/4)	41 (1-1/2)	53 (2)	63 (2-1/2)	78 (3)	91 (3-1/2)	103 (4)
THHN, THWN, THWN-2	600	0	0	0	1	1	1	2	3	5	6
	700	0	0	0	0	1	1	1	3	4	6
	750	0	0	0	0	1	1	1	3	4	5
	1000	0	0	0	0	0	1	1	1	3	4
XHHW, XHHW-2	8	3	5	8	13	19	33	50	71	97	127
	6	2	4	6	9	14	24	37	53	72	95
	4	1	3	4	7	10	18	27	38	52	69
	2	1	1	3	5	7	13	19	28	38	49
	1	1	1	2	3	5	9	14	21	28	37
	1/0	1	1	1	3	4	8	12	17	24	31
	2/0	1	1	1	2	4	7	10	15	20	26
	3/0	0	1	1	1	3	5	8	12	17	22
	4/0	0	0	1	1	2	4	7	10	14	18

250	0	1	1	1	4	5	8	11	14
300	0	0	1	1	3	5	7	9	12
350	0	0	1	1	3	4	6	8	11
400	0	0	0	1	2	4	5	7	10
500	0	0	0	1	1	3	4	6	8
600	0	0	0	1	1	2	3	5	6
700	0	0	0	0	1	1	3	4	6
750	0	0	0	1	1	1	3	4	5
1000	0	0	0	0	1	1	2	3	4

Note: Compact stranding is defined as the result of a manufacturing process where the standard conductor is compressed to the extent that the interstices (voids between strand wires) are virtually eliminated.

Source: NFPA 70, *National Electrical Code®*, NFPA, Quincy, MA, 2002 edition, Table C3(A).

Exhibit VII.45 Maximum Number of Same Size Conductors or Fixture Wires in Intermediate Metal Conduit (IMC)

		Conductors								
		Metric Designator (Trade Size)								
Type	Conductor Size (AWG/kcmil)	16 (1/2)	21 (3/4)	27 (1)	35 (1-1/4)	41 (1-1/2)	53 (2)	63 (2-1/2)	78 (3)	91 (3-1/2)	103 (4)
RHH, RHW, RHW-2	14	4	8	13	22	30	49	70	108	144	186
	12	4	6	11	18	25	41	58	89	120	154
RHH, RHW, RHW-2	10	3	5	8	15	20	33	47	72	97	124
	8	1	3	4	8	10	17	24	38	50	65
	6	1	1	3	6	8	14	19	30	40	52
	4	1	1	3	5	6	11	15	23	31	41
	3	1	1	2	4	6	9	13	21	28	36
	2	1	1	1	3	5	8	11	18	24	31
	1	0	1	1	2	3	5	7	12	16	20
	1/0	0	1	1	1	3	4	6	10	14	18
	2/0	0	1	1	1	2	4	6	9	12	15
	3/0	0	0	1	1	1	3	5	7	10	13
	4/0	0	0	1	1	1	3	4	6	9	11
	250	0	0	1	1	1	1	3	5	6	8
	300	0	0	0	1	1	1	3	4	6	7

Type	Size										
	350	0	0	0	1	1	1	2	4	5	7
	400	0	0	0	1	1	1	2	3	5	6
	500	0	0	0	1	1	1	1	3	4	5
	600	0	0	0	0	1	1	1	2	3	4
	700	0	0	0	0	1	1	1	2	3	4
	750	0	0	0	0	0	1	1	1	3	3
	800	0	0	0	0	0	1	1	1	2	3
	900	0	0	0	0	0	1	1	1	2	3
	1000	0	0	0	0	0	1	1	1	2	3
	1250	0	0	0	0	0	0	1	1	1	2
	1500	0	0	0	0	0	0	1	1	1	1
	1750	0	0	0	0	0	0	1	1	1	1
	2000	0	0	0	0	0	0	1	1	1	1
TW, THHW, THW, THW-2	14	10	17	27	47	64	104	147	228	304	392
	12	7	13	21	36	49	80	113	175	234	301
	10	5	9	15	27	36	59	84	130	174	224
	8	3	5	8	15	20	33	47	72	97	124
RHH,[a] RHW,[a] RHW-2[a]	14	6	11	18	31	42	69	98	151	202	261

Continued

Exhibit VII.45 Maximum Number of Same Size Conductors or Fixture Wires in Intermediate Metal Conduit (IMC) *Continued*

Type	Conductor Size (AWG/kcmil)	Metric Designator (Trade Size)									
		16 (1/2)	21 (3/4)	27 (1)	35 (1-1/4)	41 (1-1/2)	53 (2)	63 (2-1/2)	78 (3)	91 (3-1/2)	103 (4)
RHH,[a] RHW,[a] RHW-2[a]	12	5	9	14	25	34	56	79	122	163	209
	10	4	7	11	19	26	43	61	95	127	163
RHH,[a] RHW,[a] RHW-2[a]	8	2	4	7	12	16	26	37	57	76	98
RHH,[a] RHW,[a] RHW-2[a]	6	1	3	5	9	12	20	28	43	58	75
	4	1	2	4	6	9	15	21	32	43	56
TW, THW, THHW, THW-2	3	1	1	3	6	8	13	18	28	37	48
	2	1	1	3	5	6	11	15	23	31	41
	1	1	1	1	3	4	7	11	16	22	28
	1/0	1	1	1	3	4	6	9	14	19	24
	2/0	0	1	1	2	3	5	8	12	16	20
	3/0	0	1	1	1	3	4	6	10	13	17
	4/0	0	0	1	1	2	4	5	8	11	14

Size										
250	0	0	1	1	1	3	4	7	9	12
300	0	0	1	1	1	2	4	6	8	10
350	0	0	1	1	1	2	3	5	7	9
400	0	0	0	1	1	1	3	4	6	8
500	0	0	0	1	1	1	2	4	5	7
600	0	0	0	1	1	1	1	3	4	5
700	0	0	0	0	1	1	1	3	4	5
750	0	0	0	0	1	1	1	2	3	4
800	0	0	0	0	1	1	1	2	3	4
900	0	0	0	0	1	1	1	2	3	4
1000	0	0	0	0	0	1	1	1	3	3
1250	0	0	0	0	0	1	1	1	1	3
1500	0	0	0	0	0	0	1	1	1	2
1750	0	0	0	0	0	0	1	1	1	1
2000	0	0	0	0	0	0	1	1	1	1
THHN, THWN, THWN-2										
14	14	24	39	68	91	149	211	326	436	562
12	10	17	29	49	67	109	154	238	318	410
10	6	11	18	31	42	68	97	150	200	258
8	3	6	10	18	24	39	56	86	115	149
6	2	4	7	13	17	28	40	62	83	107

Continued

Exhibit VII.45 Maximum Number of Same Size Conductors or Fixture Wires in Intermediate Metal Conduit (IMC) *Continued*

Type	Conductor Size (AWG/kcmil)	Metric Designator (Trade Size)									
		16 (1/2)	21 (3/4)	27 (1)	35 (1-1/4)	41 (1-1/2)	53 (2)	63 (2-1/2)	78 (3)	91 (3-1/2)	103 (4)
THHN, THWN, THWN-2	4	1	3	4	8	10	17	25	38	51	66
	3	1	2	4	6	9	15	21	32	43	56
	2	1	1	3	5	7	12	17	27	36	47
	1	1	1	2	4	5	9	13	20	27	35
	1/0	1	1	1	3	4	8	11	17	23	29
	2/0	1	1	1	3	4	6	9	14	19	24
	3/0	0	1	1	2	3	5	7	12	16	20
	4/0	0	1	1	1	2	4	6	9	13	17
	250	0	0	1	1	1	3	5	8	10	13
	300	0	0	1	1	1	3	4	7	9	12
	350	0	0	1	1	1	2	4	6	8	10
	400	0	0	0	1	1	2	3	5	7	9
	500	0	0	0	1	1	1	3	4	6	7
	600	0	0	0	1	1	1	2	3	5	6
	700	0	0	0	1	1	1	2	3	4	5

Type	Size										
	750	0	0	0	1	1	1	1	3	4	5
	800	0	0	0	0	1	1	1	3	4	5
	900	0	0	0	0	1	1	1	2	3	4
	1000	0	0	0	0	1	1	1	2	3	4
FEP, FEPB, PFA, PFAH, TFE	14	13	23	38	66	89	145	205	317	423	545
	12	10	17	28	48	65	106	150	231	309	398
	10	7	12	20	34	46	76	107	166	221	285
	8	4	7	11	19	26	43	61	95	127	163
	6	3	5	8	14	19	31	44	67	90	116
	4	1	3	5	10	13	21	30	47	63	81
	3	1	3	4	8	11	18	25	39	52	68
	2	1	2	4	6	9	15	21	32	43	56
PFA, PFAH, TFE	1	1	1	2	4	6	10	14	22	30	39
PFA, PFAH, TFE, Z	1/0	1	1	1	4	5	8	12	19	25	32
	2/0	1	1	1	3	4	7	10	15	21	27
	3/0	0	1	1	2	3	6	8	13	17	22
	4/0	0	1	1	1	3	5	7	10	14	18
Z	14	16	28	46	79	107	175	247	381	510	657
	12	11	20	32	56	76	124	175	271	362	466

Continued

Exhibit VII.45 Maximum Number of Same Size Conductors or Fixture Wires in Intermediate Metal Conduit (IMC) *Continued*

Type	Conductor Size (AWG/kcmil)	Conductors Metric Designator (Trade Size)									
		16 (1/2)	21 (3/4)	27 (1)	35 (1-1/4)	41 (1-1/2)	53 (2)	63 (2-1/2)	78 (3)	91 (3-1/2)	103 (4)
Z	10	7	12	20	34	46	76	107	166	221	285
	8	4	7	12	21	29	48	68	105	140	180
	6	3	5	9	15	20	33	47	73	98	127
	4	1	3	6	10	14	23	33	50	67	87
	3	1	2	4	7	10	17	24	37	49	63
	2	1	1	3	6	8	14	20	30	41	53
	1	1	1	3	5	7	11	16	25	33	43
XHH, XHHW, XHHW-2, ZW	14	10	17	27	47	64	104	147	228	304	392
	12	7	13	21	36	49	80	113	175	234	301
	10	5	9	15	27	36	59	84	130	174	224
	8	3	5	8	15	20	33	47	72	97	124
	6	1	4	6	11	15	24	35	53	71	92
	4	1	3	4	8	11	18	25	39	52	67
	3	1	2	4	7	9	15	21	33	44	56

XHH, XHHW, XHHW-2										
2	47	37	27	18	12	7	5	3	1	1
1	35	27	20	13	9	5	4	2	1	1
1/0	30	23	17	11	8	5	3	1	1	1
2/0	25	19	14	9	6	4	3	1	1	0
3/0	20	16	12	7	5	3	2	1	1	0
4/0	17	13	10	6	4	2	1	1	0	0
250	14	11	8	5	3	1	1	1	0	0
300	12	9	7	5	3	1	1	1	0	0
350	10	8	6	4	3	1	1	1	0	0
400	9	7	5	3	2	1	1	0	0	0
500	8	6	4	3	1	1	1	0	0	0
600	6	5	3	2	1	1	1	0	0	0
700	5	4	3	1	1	1	1	0	0	0
750	5	4	3	1	1	1	1	0	0	0
800	5	4	3	1	1	1	1	0	0	0
900	4	3	2	1	1	1	0	0	0	0
1000	4	3	2	1	1	1	0	0	0	0
1250	3	2	1	1	1	1	0	0	0	0
1500	2	1	1	1	1	1	0	0	0	0
1750	2	1	1	1	1	1	0	0	0	0
2000	1	1	1	1	0	0	0	0	0	0

Continued

Exhibit VII.45 Maximum Number of Same Size Conductors or Fixture Wires in Intermediate Metal Conduit (IMC) *Continued*

Fixture Wires

Type	Conductor Size (AWG/kcmil)	Metric Designator (Trade Size)					
		16 (1/2)	21 (3/4)	27 (1)	35 (1-1/4)	41 (1-1/2)	53 (2)
FHH-2, RFH-2, RFHH-3	18	9	16	26	45	61	100
	16	8	13	22	38	51	84
SF-2, SFF-2	18	12	20	33	57	77	126
	16	10	17	27	47	64	104
	14	8	13	22	38	51	84
SF-1, SFF-1	18	21	36	59	101	137	223
RFH-1, RFHH-2, TF, TFF, XF, XFF	18	15	26	43	75	101	165
RFH-2, TF, TFF, XF, XFF	16	12	21	35	60	81	133
XF, XFF	14	10	17	27	47	64	104

TFN, TFFN	18	25	42	69	119	161	264
	16	19	32	53	91	123	201
PF, PFF, PGF, PGFF, PAF, PTF, PTFF, PAFF	18	23	40	66	113	153	250
	16	18	31	51	87	118	193
	14	13	23	38	66	89	145
ZF, ZFF, ZHF, HF, HFF	18	30	52	85	146	197	322
	16	22	38	63	108	145	238
	14	16	28	46	79	107	175
KF-2, KFF-2	18	44	75	123	212	287	468
	16	31	53	87	149	202	330
	14	21	36	60	103	139	227
	12	14	25	41	70	95	156
	10	10	17	27	47	64	104
KF-1, KFF-1	18	52	90	147	253	342	558
	16	37	63	103	178	240	392
	14	25	42	69	119	161	264
	12	16	28	46	79	107	175
	10	10	18	30	52	70	114

Continued

Exhibit VII.45 Maximum Number of Same Size Conductors or Fixture Wires in Intermediate Metal Conduit (IMC) *Continued*

Fixture Wires

Type	Conductor Size (AWG/kcmil)	Metric Designator (Trade Size)					
		16 (1/2)	21 (3/4)	27 (1)	35 (1-1/4)	41 (1-1/2)	53 (2)
XF, XFF	12	5	9	14	25	34	56
	10	4	7	11	19	26	43

Note: This table is for concentric stranded conductors only. For compact stranded conductors, Exhibit VII.46 should be used.

[a]Types RHH, RHW, and RHW-2 without outer covering.

Source: NFPA 70, *National Electrical Code®*, NFPA, Quincy, MA, 2002 edition, Table C4.

Exhibit VII.46 Maximum Number of Same Size Compact Conductors in Intermediate Metal Conduit (IMC)

	Conductor Size (AWG/kcmil)	Compact Conductors — Metric Designator (Trade Size)									
Type		16 (1/2)	21 (3/4)	27 (1)	35 (1-1/4)	41 (1-1/2)	53 (2)	63 (2-1/2)	78 (3)	91 (3-1/2)	103 (4)
THW, THW-2, THHW	8	2	4	7	13	17	28	40	62	83	107
	6	1	3	6	10	13	22	31	48	64	82
	4	1	2	4	7	10	16	23	36	48	62
	2	1	1	3	5	7	12	17	26	35	45
	1	1	1	1	4	5	8	12	18	25	32
	1/0	1	1	1	3	4	7	10	16	21	27
	2/0	0	1	1	3	4	6	9	13	18	23
	3/0	0	1	1	2	3	5	7	11	15	20
	4/0	0	1	1	1	2	4	6	9	13	16
	250	0	0	1	1	1	3	5	7	10	13
	300	0	0	1	1	1	3	4	6	9	11
	350	0	0	1	1	1	2	4	6	8	10

Continued

Exhibit VII.46 Maximum Number of Same Size Compact Conductors in Intermediate Metal Conduit (IMC) (Continued)

Type	Conductor Size (AWG/kcmil)	Compact Conductors — Metric Designator (Trade Size)									
		16 (1/2)	21 (3/4)	27 (1)	35 (1-1/4)	41 (1-1/2)	53 (2)	63 (2-1/2)	78 (3)	91 (3-1/2)	103 (4)
THW, THW-2, THHW	400	0	0	1	1	1	2	3	5	7	9
	500	0	0	1	1	1	1	3	4	6	8
	600	0	0	0	1	1	1	2	3	5	6
	700	0	0	0	1	1	1	1	3	4	5
	750	0	0	0	1	1	1	1	3	4	5
	1000	0	0	0	0	1	1	1	2	3	4
THHN, THWN, THWN-2	8	—	5	8	14	19	32	45	70	93	120
	6	3	3	5	9	12	20	28	43	58	74
	4	1	1	3	6	8	14	20	31	41	53
	2	1	1	3	5	6	10	15	23	31	40
	1	1	1	3	4	5	7	10	16	22	28
	1/0	1	1	2	4	5	9	13	20	26	34
	2/0	1	1	1	3	4	7	10	16	22	28

	24	18	14	9	6	4	3	1	1	0
3/0	24	18	14	9	6	4	3	1	1	0
4/0	19	15	11	7	5	3	2	1	1	0
250	15	12	9	6	4	2	1	1	1	0
300	13	10	7	5	3	1	1	1	1	0
350	11	9	7	4	3	1	1	1	1	0
400	10	8	6	4	2	1	1	1	1	0
500	9	7	5	3	2	1	1	1	1	0
600	7	5	4	2	1	1	1	0	0	0
700	6	5	3	2	1	1	1	0	0	0
750	6	4	3	1	1	1	0	0	0	0
1000	4	3	2	1	1	1	0	0	0	0
XHHW, XHHW-2										
8	138	107	80	52	37	22	16	9	6	3
6	103	80	59	38	27	16	12	7	4	2
4	74	58	43	28	20	12	9	5	3	1
2	53	41	31	20	14	8	6	3	1	1
1	40	31	23	15	10	6	5	3	1	1
1/0	34	26	20	13	9	5	4	2	1	1
2/0	29	22	17	11	7	4	3	1	1	1

Continued

Exhibit VII.46 Maximum Number of Same Size Compact Conductors in Intermediate Metal Conduit (IMC) *(Continued)*

	Conductor Size (AWG/kcmil)	Metric Designator (Trade Size)									
Type		16 (1/2)	21 (3/4)	27 (1)	35 (1-1/4)	41 (1-1/2)	53 (2)	63 (2-1/2)	78 (3)	91 (3-1/2)	103 (4)
XHHW, XHHW-2	3/0	0	1	1	3	4	6	9	14	18	24
	4/0	0	1	1	2	3	5	7	11	15	20
	250	0	1	1	1	2	4	6	9	12	16
	300	0	0	1	1	1	3	5	8	10	13
	350	0	0	1	1	1	3	4	7	9	12
	400	0	0	1	1	1	3	4	6	8	11
	500	0	0	1	1	1	2	3	5	7	9
	600	0	0	0	1	1	1	2	4	5	7
	700	0	0	0	1	1	1	2	3	5	6
	750	0	0	0	1	1	1	1	3	4	6
	1000	0	0	0	0	1	1	1	2	3	4

Note: Compact stranding is defined as the result of a manufacturing process where the standard conductor is compressed to the extent that the interstices (voids between strand wires) are virtually eliminated.

Source: NFPA 70, *National Electrical Code®*, NFPA, Quincy, MA, 2002 edition, Table C4(A).

Exhibit VII.47 Maximum Number of Same Size Conductors or Fixture Wires in Liquidtight Flexible Nonmetallic Conduit (Type LFNC-B[a])

Type	Conductor Size (AWG/kcmil)	Metric Designator (Trade Size)						
		12 (3/8)	16 (1/2)	21 (3/4)	27 (1)	35 (1-1/4)	41 (1-1/2)	53 (2)
RHH, RHW, RHW-2	14	2	4	7	12	21	27	44
	12	1	3	6	10	17	22	36
RHH, RHW, RHW-2	10	1	3	5	8	14	18	29
	8	1	1	2	4	7	9	15
	6	1	1	1	3	6	7	12
	4	0	1	1	2	4	6	9
	3	0	1	1	1	4	5	8
	2	0	1	1	1	3	4	7
	1	0	0	1	1	1	3	5
	1/0	0	0	1	1	1	2	4
	2/0	0	0	1	1	1	1	3
	3/0	0	0	0	1	1	1	3
	4/0	0	0	0	1	1	1	2

Continued

Exhibit VII.47 Maximum Number of Same Size Conductors or Fixture Wires in Liquidtight Flexible Nonmetallic Conduit (Type LFNC-B*) *Continued*

		Conductors						
		Metric Designator (Trade Size)						
Type	Conductor Size (AWG/kcmil)	12 (3/8)	16 (1/2)	21 (3/4)	27 (1)	35 (1-1/4)	41 (1-1/2)	53 (2)
RHH, RHW, RHW-2	250	0	0	0	0	1	1	1
	300	0	0	0	0	1	1	1
	350	0	0	0	0	1	1	1
	400	0	0	0	0	1	1	1
	500	0	0	0	0	1	1	1
	600	0	0	0	0	0	1	1
	700	0	0	0	0	0	1	1
	750	0	0	0	0	0	1	1
	800	0	0	0	0	0	1	1
	900	0	0	0	0	0	0	1
	1000	0	0	0	0	0	0	0
	1250	0	0	0	0	0	0	0
	1500	0	0	0	0	0	0	0
	1750	0	0	0	0	0	0	0
	2000	0	0	0	0	0	0	0

Type	Size							
TW, THHW, THW, THW-2	14	5	9	15	25	44	57	93
	12	4	7	12	19	33	43	71
	10	3	5	9	14	25	32	53
	8	1	3	5	8	14	18	29
RHH,[b] RHW,[b] RHW-2[b]	14	3	6	10	16	29	38	62
RHH,[b] RHW,[b] RHW-2[b]	12	3	5	8	13	23	30	50
	10	1	3	6	10	18	23	39
RHH,[b] RHW,[b] RHW-2[b]	8	1	1	4	6	11	14	23
RHH,[b] RHW,[b] RHW-2,[b] TW, THW	6	1	1	3	5	8	11	18
	4	1	1	1	3	6	8	13
	3	1	1	1	3	5	7	11
THHW, THW-2	2	0	1	1	2	4	6	9
	1	0	1	1	1	3	4	7
	1/0	0	0	1	1	2	3	6
	2/0	0	0	1	1	2	3	5
	3/0	0	0	1	1	1	2	4
	4/0	0	0	0	1	1	1	3

Continued

Exhibit VII.47 Maximum Number of Same Size Conductors or Fixture Wires in Liquidtight Flexible Nonmetallic Conduit (Type LFNC-B*) *Continued*

Type	Conductor Size (AWG/kcmil)	Metric Designator (Trade Size)						
		12 (3/8)	16 (1/2)	21 (3/4)	27 (1)	35 (1-1/4)	41 (1-1/2)	53 (2)
THHW, THW-2	250	0	0	0	1	1	1	3
	300	0	0	0	1	1	1	2
	350	0	0	0	0	1	1	1
	400	0	0	0	0	1	1	1
	500	0	0	0	0	1	1	1
	600	0	0	0	0	1	1	1
	700	0	0	0	0	0	1	1
	750	0	0	0	0	0	1	1
	800	0	0	0	0	0	1	1
	900	0	0	0	0	0	1	1
	1000	0	0	0	0	0	0	1
	1250	0	0	0	0	0	0	1
	1500	0	0	0	0	0	0	0

THHN, THWN, THWN-2

Conductor Size							
1750	0	0	0	0	0	0	0
2000	0	0	0	0	0	0	0
14	133	81	63	36	22	13	8
12	97	59	46	26	16	9	5
10	61	37	29	16	10	6	3
8	35	21	16	9	6	3	1
6	25	15	12	7	4	2	1
4	15	9	7	4	2	1	1
3	13	8	6	3	1	1	1
2	11	7	5	3	1	1	1
1	8	5	4	1	1	1	0
1/0	7	4	3	1	1	1	0
2/0	6	3	2	1	1	0	0
3/0	5	3	1	1	1	0	0
4/0	4	2	1	1	1	0	0
250	3	1	1	1	0	0	0
300	3	1	1	1	0	0	0
350	2	1	1	1	0	0	0
400	1	1	1	0	0	0	0
500	1	1	1	0	0	0	0

Continued

Exhibit VII.47 Maximum Number of Same Size Conductors or Fixture Wires in Liquidtight Flexible Nonmetallic Conduit (Type LFNC-B°) *Continued*

| | | Conductors | | | | | | |
| | | Metric Designator (Trade Size) | | | | | | |
Type	Conductor Size (AWG/kcmil)	12 (3/8)	16 (1/2)	21 (3/4)	27 (1)	35 (1-1/4)	41 (1-1/2)	53 (2)
THHN, THWN, THWN-2	600	0	0	0	0	1	1	1
	700	0	0	0	0	1	1	1
	750	0	0	0	0	0	1	1
	800	0	0	0	0	0	1	1
	900	0	0	0	0	0	1	1
	1000	0	0	0	0	0	0	1
FEP, FEPB, PFA, PFAH, TFE	14	7	12	21	35	61	79	129
	12	5	9	15	25	44	57	94
	10	4	6	11	18	32	41	68
	8	1	3	6	10	18	23	39
	6	1	2	4	7	13	17	27
	4	1	1	3	5	9	12	19
	3	1	1	2	4	7	10	16
	2	1	1	1	3	6	8	13

Type	Size	0	1	1	2	4	5	9
PFA, PFAH, TFE	1	0	1	1	1	3	4	9
PFA, PFAH	1/0	0	1	1	1	3	4	7
TFE, Z	2/0	0	1	1	1	3	4	6
	3/0	0	0	1	1	2	3	5
	4/0	0	0	1	1	1	2	4
Z	14	9	15	26	42	73	95	156
	12	6	10	18	30	52	67	111
	10	4	6	11	18	32	41	68
	8	2	4	7	11	20	26	43
	6	1	3	5	8	14	18	30
	4	1	1	3	5	9	12	20
	3	1	1	2	4	7	9	15
	2	0	1	1	3	6	7	12
	1	0	1	1	2	5	6	10
XHH, XHHW, XHHW-2, ZW	14	5	9	15	25	44	57	93
	12	4	7	12	19	33	43	71
	10	3	5	9	14	25	32	53
	8	1	3	5	8	14	18	29
	6	1	1	3	6	10	13	22

Continued

Exhibit VII.47 Maximum Number of Same Size Conductors or Fixture Wires in Liquidtight Flexible Nonmetallic Conduit (Type LFNC-B^a) *Continued*

Type	Conductor Size (AWG/kcmil)	Conductors						
		Metric Designator (Trade Size)						
		12 (3/8)	16 (1/2)	21 (3/4)	27 (1)	35 (1-1/4)	41 (1-1/2)	53 (2)
XHH, XHHW, XHHW-2, ZW	4	1	1	2	4	7	9	16
	3	1	1	1	3	6	8	13
	2	1	1	1	3	5	7	11
XHH, XHHW, XHHW-2	1	0	1	1	1	4	5	8
	1/0	0	1	1	1	3	4	7
	2/0	0	0	1	1	2	3	6
	3/0	0	0	1	1	1	3	5
	4/0	0	0	0	1	1	2	4
	250	0	0	0	1	1	1	3
	300	0	0	0	1	1	1	3
	350	0	0	0	1	1	1	2
	400	0	0	0	0	1	1	1
	500	0	0	0	0	0	1	1

Type	Conductor Size (AWG/kcmil)	Metric Designator (Trade Size)						
		12 (3/8)	16 (1/2)	21 (3/4)	27 (1)	35 (1-1/4)	41 (1-1/2)	53 (2)
	600	0	0	0	0	1	1	1
	700	0	0	0	0	1	1	1
	750	0	0	0	0	0	1	1
	800	0	0	0	0	0	1	1
	900	0	0	0	0	0	0	1
	1000	0	0	0	0	0	0	0
	1250	0	0	0	0	0	0	0
	1500	0	0	0	0	0	0	0
	1750	0	0	0	0	0	0	0
	2000	0	0	0	0	0	0	0

Fixture Wires

Type	Conductor Size (AWG/kcmil)	Metric Designator (Trade Size)						
		12 (3/8)	16 (1/2)	21 (3/4)	27 (1)	35 (1-1/4)	41 (1-1/2)	53 (2)
FFH-2, RFH-2, SF-2, SFF-2	18	5	8	15	24	42	54	89
	16	4	7	12	20	35	46	75

Continued

Exhibit VII.47 Maximum Number of Same Size Conductors or Fixture Wires in Liquidtight Flexible Nonmetallic Conduit (Type LFNC-B^a) *Continued*

Fixture Wires

Type	Conductor Size (AWG/kcmil)	12 (3/8)	16 (1/2)	21 (3/4)	27 (1)	35 (1-1/4)	41 (1-1/2)	53 (2)
					Metric Designator (Trade Size)			
FFH-2, RFH-2, SF-2, SFF-2	18	6	11	19	30	53	69	113
	16	5	9	15	25	44	57	93
	14	4	7	12	20	35	46	75
SF-1, SFF-1	18	11	19	33	53	94	122	199
RFH-1, RFHH-2, TF, TFF, XF, XFF	18	8	14	24	39	69	90	147
RFHH-2, TF, TFF, XF, XFF	16	7	11	20	32	56	72	119
XF, XFF	14	5	9	15	25	44	57	93
TFN, TFFN	18	14	23	39	63	111	144	236
	16	10	17	30	48	85	110	180
PF, PFF, PGF, PGFF, PAF, PTF, PTFF, PAFF	18	13	21	37	60	105	136	223
	16	10	16	29	46	81	105	173
	14	7	12	21	35	61	79	129

Type	Size							
HF, HFF, ZF, ZFF, ZHF	18	17	28	48	77	136	176	288
	16	12	20	35	57	100	129	212
	14	9	15	26	42	73	95	156
KF-2, KFF-2	18	24	40	70	112	197	255	418
	16	17	28	49	79	139	180	295
	14	12	19	34	54	95	123	202
	12	8	13	23	37	65	85	139
	10	5	9	15	25	44	57	93
KF-1, KFF-1	18	29	48	83	134	235	304	499
	16	20	34	58	94	165	214	350
	14	14	23	39	63	111	144	236
	12	9	15	26	42	73	95	156
	10	6	10	17	27	48	62	102
XF, XFF	12	3	5	8	13	23	30	50
	10	1	3	6	10	18	23	39

Note: This table is for concentric strandec conductors only. For compact stranded conductors, Exhibit VII.48 should be used.
[a]Corresponds to 356.2(2) of the *NEC*.
[b]Types RHH, RHW, and RHW-2 without outer covering.
Source: NFPA 70, *National Electrical Code®*, NFPA, Quincy, MA, 2002 edition, Table C5.

Exhibit VII.48 Maximum Number of Same Size Compact Conductors in Liquidtight Flexible Nonmetallic Conduit (Type LFNC-B^a)

	Conductor Size (AWG/kcmil)	Compact Conductors — Metric Designator (Trade Size)						
Type		12 (3/8)	16 (1/2)	21 (3/4)	27 (1)	35 (1-1/4)	41 (1-1/2)	53 (2)
THW, THW-2, THHW	8	1	2	4	7	12	15	25
	6	1	1	3	5	9	12	19
	4	1	1	2	4	7	9	14
	2	1	1	1	3	5	6	11
	1	0	1	1	1	3	4	7
	1/0	0	0	1	1	3	4	6
	2/0	0	0	1	1	2	3	5
	3/0	0	0	1	1	1	3	4
	4/0	0	0	1	1	1	2	4
	250	0	0	0	1	1	1	3
	300	0	0	0	1	1	1	2
	350	0	0	0	1	1	1	2

Size							
400	1	1	1	0	0	0	0
500	1	1	1	0	0	0	0
600	1	1	1	0	0	0	0
700	1	1	0	0	0	0	0
750	1	1	0	0	0	0	0
1000	1	1	0	0	0	0	0
THHN, THWN, THWN-2							
8	—	—	—	—	—	—	—
6	28	17	13	7	4	2	1
4	17	11	8	4	3	1	1
2	12	7	6	3	1	1	1
1	9	6	4	2	1	1	0
1/0	8	5	4	1	1	1	0
2/0	6	4	3	1	1	1	0
3/0	5	3	2	1	1	0	0
4/0	4	3	1	1	1	0	0
250	3	2	2	1	1	0	0
300	3	1	1	1	0	0	0
350	2	1	1	1	0	0	0
400	2	1	1	0	0	0	0
500	1	1	1	0	0	0	0

Continued

Exhibit VII.48 Maximum Number of Same Size Compact Conductors in Liquidtight Flexible Nonmetallic Conduit (Type LFNC-B[a]) *Continued*

Type	Conductor Size (AWG/kcmil)	Compact Conductors						
		Metric Designator (Trade Size)						
		12 (3/8)	16 (1/2)	21 (3/4)	27 (1)	35 (1-1/4)	41 (1-1/2)	53 (2)
THHN, THWN, THWN-2	600	0	0	0	0	1	1	1
	700	0	0	0	0	1	1	1
	750	0	0	0	0	1	1	1
	1000	0	0	0	0	0	1	1
XHHW, XHHW-2	8	1	3	5	9	15	20	33
	6	1	2	4	6	11	15	24
	4	1	1	3	4	8	11	17
	2	1	1	1	3	6	7	12
	1	0	1	1	2	4	6	9
	1/0	0	1	1	1	4	5	8
	2/0	0	1	1	1	3	4	7
	3/0	0	0	1	1	2	3	5
	4/0	0	0	1	1	1	3	4

250	0	0	1	1	1	3
300	0	0	0	1	1	3
350	0	0	0	1	1	3
400	0	0	0	1	1	2
500	0	0	0	1	1	1
600	0	0	0	0	1	1
700	0	0	0	0	1	1
750	0	0	0	0	1	1
1000	0	0	0	0	0	1

[a]Corresponds to 356.2(2) of the *NEC*.

Note: Compact stranding is defined as the result of a manufacturing process where the standard conductor is compressed to the extent that the interstices (voids between strand wires) are virtually eliminated.

Source: NFPA 70, National Electrical Code®, NFPA, Quincy, MA, 2002 edition, Table C5(A).

Exhibit VII.49 Maximum Number of Same Size Conductors or Fixture Wires in Liquidtight Flexible Nonmetallic Conduit (Type LFNC-A[a])

Type	Conductor Size (AWG/kcmil)	Conductors Metric Designator (Trade Size)						
		12 (3/8)	16 (1/2)	21 (3/4)	27 (1)	35 (1-1/4)	41 (1-1/2)	53 (2)
RHH, RHW, RHW-2	14	2	4	7	11	20	27	45
	12	1	3	6	9	17	23	38
	10	1	3	5	8	13	18	30
	8	1	1	2	4	7	9	16
	6	1	1	1	3	5	7	13
	4	0	1	1	2	4	6	10
	3	0	1	1	1	4	5	8
	2	0	0	1	1	3	4	7
	1	0	0	1	1	1	3	5
	1/0	0	0	1	1	1	2	4
	2/0	0	0	1	1	1	1	4
	3/0	0	0	0	1	1	1	3
	4/0	0	0	0	1	1	1	3

Size							
250	0	0	0	0	1	1	1
300	0	0	0	0	1	1	1
350	0	0	0	0	1	1	1
400	0	0	0	0	1	1	1
500	0	0	0	0	0	1	1
600	0	0	0	0	0	1	1
700	0	0	0	0	0	0	1
750	0	0	0	0	0	0	1
800	0	0	0	0	0	0	1
900	0	0	0	0	0	0	1
1000	0	0	0	0	0	0	1
1250	0	0	0	0	0	0	0
1500	0	0	0	0	0	0	0
1750	0	0	0	0	0	0	0
2000	0	0	0	0	0	0	0
14	5	9	15	24	43	58	96
12	4	7	12	19	33	44	74
10	3	5	9	14	24	33	55
8	1	3	5	8	13	18	30

TW, THHW, THW, THW-2

Continued

Exhibit VII.49 Maximum Number of Same Size Conductors or Fixture Wires in Liquidtight Flexible Nonmetallic Conduit (Type LFNC-A^a) *Continued*

Type	Conductor Size (AWG/kcmil)	Metric Designator (Trade Size)						
		12 (3/8)	16 (1/2)	21 (3/4)	27 (1)	35 (1-1/4)	41 (1-1/2)	53 (2)
RHH,^{b} RHW,^{b} RHW-2^b	14	3	6	10	16	28	38	64
	12	3	4	8	13	23	31	51
	10	2	3	6	10	18	24	40
	8	1	1	4	6	10	14	24
RHH,^{b} RHW,^{b} RHW-2,^{b} TW, THW, THHW, THW-2	6	1	1	3	4	8	11	18
	4	1	1	1	3	6	8	13
	3	1	1	1	3	5	7	11
	2	0	1	1	2	4	6	10
	1	0	1	1	1	3	4	7
	1/0	0	0	1	1	2	3	6
	2/0	0	0	1	1	2	3	5
	3/0	0	0	1	1	1	2	4
	4/0	0	0	0	1	1	1	3

Continued

THHN, THWN, THWN-2

Size							
250	3	1	1	1	0	0	0
300	2	1	1	1	0	0	0
350	1	1	1	0	0	0	0
400	1	1	1	0	0	0	0
500	1	1	1	0	0	0	0
600	1	1	1	0	0	0	0
700	1	1	0	0	0	0	0
750	1	1	0	0	0	0	0
800	1	1	0	0	0	0	0
900	1	0	0	0	0	0	0
1000	1	0	0	0	0	0	0
1250	1	0	0	0	0	0	0
1500	1	0	0	0	0	0	0
1750	0	0	0	0	0	0	0
2000	0	0	0	0	0	0	0
14	137	83	62	35	22	13	8
12	100	60	45	25	16	9	5

Exhibit VII.49 Maximum Number of Same Size Conductors or Fixture Wires in Liquidtight Flexible Nonmetallic Conduit (Type LFNC-A*) *Continued*

	Conductors						
Conductor Size (AWG/kcmil)	Metric Designator (Trade Size)						
	12 (3/8)	16 (1/2)	21 (3/4)	27 (1)	35 (1-1/4)	41 (1-1/2)	53 (2)
Type THHN, THWN, THWN-2							
10	3	6	10	16	28	38	63
8	1	3	6	9	16	22	36
6	1	2	4	6	12	16	26
4	1	1	2	4	7	9	16
3	1	1	1	3	6	8	13
2	1	1	1	3	5	7	11
1	0	1	1	1	4	5	8
1/0	0	0	1	1	3	4	7
2/0	0	0	1	1	2	3	6
3/0	0	0	1	1	1	3	5
4/0	0	0	1	1	1	2	4
250	0	0	0	1	1	1	3
300	0	0	0	1	1	1	3
350	0	0	0	1	1	1	2

400	0	0	0	0	1	1	1
500	0	0	0	0	1	1	1
600	0	0	0	0	1	1	1
700	0	0	0	0	1	1	1
750	0	0	0	0	0	1	1
800	0	0	0	0	0	1	1
900	0	0	0	0	0	1	1
1000	0	0	0	0	0	0	1
FEP, FEPB, PFA, PFAH, TFE							
14	7	12	21	34	60	80	133
12	5	9	15	25	44	59	97
10	4	6	11	18	31	42	70
8	1	3	6	10	18	24	40
6	1	2	4	7	13	17	28
PFA, PFAH, TFE							
4	1	1	3	5	9	12	20
3	1	1	2	4	7	10	16
2	1	1	1	3	6	8	13
1	0	1	1	2	4	5	9
PFA, PFAH, TFE, Z							
1/0	0	1	1	1	3	5	8
2/0	0	1	1	1	3	4	6

Continued

Exhibit VII.49 Maximum Number of Same Size Conductors or Fixture Wires in Liquidtight Flexible Nonmetallic Conduit (Type LFNC-A⁽ᵃ⁾) *Continued*

Type	Conductor Size (AWG/kcmil)	Conductors 12 (3/8)	16 (1/2)	Metric Designator (Trade Size) 21 (3/4)	27 (1)	35 (1-1/4)	41 (1-1/2)	53 (2)
PFA, PFAH, TFE, Z	3/0	0	0	1	1	2	3	5
	4/0	0	0	1	1	1	2	4
Z	14	9	15	25	41	72	97	161
	12	6	10	18	29	51	69	114
	10	4	6	11	18	31	42	70
	8	2	4	7	11	20	26	44
	6	1	3	5	8	14	18	31
	4	1	1	3	5	9	13	21
	3	1	1	2	4	7	9	15
	2	1	1	1	3	6	8	13
	1	1	1	1	2	4	6	10
XHH, XHHW, XHHW-2, ZW	14	5	9	15	24	43	58	96
	12	4	7	12	19	33	44	74
	10	3	5	9	14	24	33	55

Continued

XHH, XHHW, XHHW-2							
8	1	3	5	8	13	18	30
6	1	1	3	5	10	13	22
4	1	1	2	4	7	10	16
3	1	1	1	3	6	8	14
2	1	1	1	3	5	7	11
1	0	1	1	3	4	5	8
1/0	0	1	1	1	3	4	7
2/0	0	0	1	1	2	3	6
3/0	0	0	1	1	1	3	5
4/0	0	0	1	1	1	2	4
250	0	0	0	1	1	1	3
300	0	0	0	1	1	1	3
350	0	0	0	0	1	1	2
400	0	0	0	0	1	1	1
500	0	0	0	0	1	1	1
600	0	0	0	0	1	1	1
700	0	0	0	0	0	1	1
750	0	0	0	0	0	1	1
800	0	0	0	0	0	1	1
900	0	0	0	0	0	1	1

Exhibit VII.49 Maximum Number of Same Size Conductors or Fixture Wires in Liquidtight Flexible Nonmetallic Conduit (Type LFNC-A³) *Continued*

Conductors

Type	Conductor Size (AWG/kcmil)	Metric Designator (Trade Size)						
		12 (3/8)	16 (1/2)	21 (3/4)	27 (1)	35 (1-1/4)	41 (1-1/2)	53 (2)
XHH, XHHW, XHHW-2	1000	0	0	0	0	0	0	1
	1250	0	0	0	0	0	0	1
	1500	0	0	0	0	0	0	1
	1750	0	0	0	0	0	0	0
	2000	0	0	0	0	0	0	0

Fixture Wires

Type	Conductor Size (AWG/kcmil)	Metric Designator (Trade Size)						
		12 (3/8)	16 (1/2)	21 (3/4)	27 (1)	35 (1-1/4)	41 (1-1/2)	53 (2)
FFH-2, RFH-2, RFHH-3	18	5	8	14	23	41	55	92
	16	4	7	12	20	35	47	77
	16	4	7	12	20	35	47	77

Type	Size							
SF-2, SFF-2	18	6	11	18	29	52	70	116
	16	5	9	15	24	43	58	96
	14	4	7	12	20	35	47	77
SF-1, SFF-1	18	12	19	33	52	92	124	205
RFH-1, RFHH-2, TF, TFF, XF, XFF	18	8	14	24	39	68	91	152
RFHH-2, TF, TFF, XF, XFF	16	7	11	19	31	55	74	122
XF, XFF	14	5	9	15	24	43	58	96
TFN, TFFN	18	14	22	39	62	109	146	243
	16	10	17	29	47	83	112	185
PF, PFF, PGF, PGFF, PAF, PTF, PTFF, PAFF	18	13	21	37	59	103	139	230
	16	10	16	28	45	80	107	178
	14	7	12	21	34	60	80	161
HF, HFF, ZF, ZFF, ZHF	18	17	27	47	76	133	179	297
	16	12	20	35	56	98	132	219
	14	9	15	25	41	72	97	161
KF-2, KFF-2	18	25	40	69	110	193	260	431
	16	17	28	48	77	136	183	303
	14	12	19	33	53	94	126	209

Continued

Exhibit VII.49 Maximum Number of Same Size Conductors or Fixture Wires in Liquidtight Flexible Nonmetallic Conduit (Type LFNC-A[a]) *Continued*

Fixture Wires

Type	Conductor Size (AWG/kcmil)	Metric Designator (Trade Size)						
		12 (3/8)	16 (1/2)	21 (3/4)	27 (1)	35 (1-1/4)	41 (1-1/2)	53 (2)
KF-2, KFF-2	12	8	13	23	36	64	86	143
	10	5	9	15	24	43	58	96
KF-1, KFF-1	18	29	48	82	131	231	310	514
	16	21	33	57	92	162	218	361
	14	14	22	39	62	109	146	243
	12	9	15	25	41	72	97	161
	10	6	10	17	27	47	63	105
XF, XFF	12	3	4	8	13	23	31	51
	10	1	3	6	10	18	24	40

Note: This table is for concentric standard conductors only. For compact stranded conductors, Exhibit VII.50 should be used.

[a]Corresponds to 356.2(1) of the *NEC*.

[b]Types RHH, RHW, and RHW-2 without outer covering.

Source: NFPA 70, *National Electrical Code®*, NFPA, Quincy, MA, 2002 edition, Table C6.

Exhibit VII.50 Maximum Number of Same Size Compact Conductors in Liquidtight Flexible Nonmetallic Conduit (Type LFNC-A[a])

Type	Conductor Size (AWG/kcmil)	Compact Conductors Metric Designator (Trade Size)						
		12 (3/8)	16 (1/2)	21 (3/4)	27 (1)	35 (1-1/4)	41 (1-1/2)	53 (2)
THW, THW-2, THHW	8	1	2	4	6	11	16	26
	6	1	1	3	5	9	12	20
	4	1	1	2	4	7	9	15
	2	1	1	1	3	5	6	11
	1	0	1	1	1	3	4	8
	1/0	0	0	1	1	3	4	7
	2/0	0	0	1	1	2	3	5
	3/0	0	0	1	1	1	3	5
	4/0	0	0	1	1	1	2	4
	250	0	0	0	1	1	1	3
	300	0	0	0	1	1	1	3
	350	0	0	0	1	1	1	2

Continued

Exhibit VII.50 Maximum Number of Same Size Compact Conductors in Liquidtight Flexible Nonmetallic Conduit (Type LFNC-A⁹) Continued

| | | Compact Conductors | | | | | | |
| | | Metric Designator (Trade Size) | | | | | | |
Type	Conductor Size (AWG/kcmil)	12 (3/8)	16 (1/2)	21 (3/4)	27 (1)	35 (1-1/4)	41 (1-1/2)	53 (2)
THW, THW-2, THHW	400	0	0	0	0	1	1	1
	500	0	0	0	0	1	1	1
	600	0	0	0	0	1	1	1
	700	0	0	0	0	1	1	1
	750	0	0	0	0	0	1	1
	1000	0	0	0	0	0	1	1
THHN, THWN, THWN-2	8	—	—	—	—	—	—	—
	6	1	2	4	7	13	18	29
	4	1	1	3	4	8	11	18
	2	1	1	1	3	6	8	13
	1	0	1	1	2	4	6	10
	1/0	0	1	1	1	3	5	8
	2/0	0	1	1	1	3	4	7

Type	Size							
	3/0	0	0	1	1	1	2	3	6
	4/0	0	0	1	1	1	1	3	5
	250	0	0	1	1	1	1	3
	300	0	0	0	1	1	1	3
	350	0	0	0	1	1	1	3
	400	0	0	0	1	1	1	2
	500	0	0	0	0	1	1	1
	600	0	0	0	0	1	1	1
	700	0	0	0	0	1	1	1
	750	0	0	0	0	1	1	1
	1000	0	0	0	0	0	1	1
XHHW, XHHW-2	8	1	3	5	8	15	20	34
	6	1	2	4	6	11	15	25
	4	1	1	3	4	8	11	18
	2	0	1	1	3	6	8	13
	1	0	1	1	2	4	6	10
	1/0	0	1	1	1	3	5	8
	2/0	0	1	1	1	3	4	7
	3/0	0	0	1	1	2	3	6
	4/0	0	0	1	1	1	3	5

Continued

Exhibit VII.50 Maximum Number of Same Size Compact Conductors in Liquidtight Flexible Nonmetallic Conduit (Type LFNC-A[a]) *Continued*

Type	Conductor Size (AWG/kcmil)	Compact Conductors						
		Metric Designator (Trade Size)						
		12 (3/8)	16 (1/2)	21 (3/4)	27 (1)	35 (1-1/4)	41 (1-1/2)	53 (2)
XHHW, XHHW-2	250	0	0	1	1	1	2	4
	300	0	0	0	1	1	1	3
	350	0	0	0	1	1	1	3
	400	0	0	0	1	1	1	2
	500	0	0	0	0	1	1	1
	600	0	0	0	0	1	1	1
	700	0	0	0	0	1	1	1
	750	0	0	0	0	1	1	1
	1000	0	0	0	0	0	1	1

[a]Corresponds to 356.2(1) of the NEC.

Note: Compact stranding is defined as the result of a manufacturing process where the standard conductor is compressed to the extent that the interstices (voids between strand wires) are virtually eliminated.

Source: NFPA 70, National Electrical Code®, NFPA, Quincy, MA, 2002 edition, Table C6(A).

Exhibit VII.51 Maximum Number of Same Size Conductors or Fixture Wires in Liquidtight Flexible Metal Conduit (LFMC)

Conductors

Type	Conductor Size (AWG/kcmil)	Metric Designator (Trade Size)									
		16 (1/2)	21 (3/4)	27 (1)	35 (1-1/4)	41 (1-1/2)	53 (2)	63 (2-1/2)	78 (3)	91 (3-1/2)	103 (4)
RHH, RHW, RHW-2	14	4	7	12	21	27	44	66	102	133	173
	12	3	6	10	17	22	36	55	84	110	144
	10	3	5	8	14	18	29	44	68	89	116
	8	1	2	4	7	9	15	23	36	46	61
	6	1	1	3	6	7	12	18	28	37	48
	4	1	1	2	4	6	9	14	22	29	38
	3	1	1	1	4	5	8	13	19	25	33
	2	1	1	1	3	4	7	11	17	22	29
	1	0	1	1	1	3	5	7	11	14	19
	1/0	0	1	1	2	2	4	6	10	13	16
	2/0	0	1	1	1	1	3	5	8	11	14
	3/0	0	0	1	1	1	3	4	7	9	12
	4/0	0	0	1	1	1	2	4	6	8	10

Continued

Exhibit VII.51 Maximum Number of Same Size Conductors or Fixture Wires in Liquidtight Flexible Metal Conduit (LFMC) *Continued*

Type	Conductor Size (AWG/kcmil)	Conductors — Metric Designator (Trade Size)									
		16 (1/2)	21 (3/4)	27 (1)	35 (1-1/4)	41 (1-1/2)	53 (2)	63 (2-1/2)	78 (3)	91 (3-1/2)	103 (4)
RHH, RHW, RHW-2	250	0	0	0	0	1	1	3	4	6	8
	300	0	0	0	1	1	1	2	4	5	7
	350	0	0	0	1	1	1	2	3	5	6
	400	0	0	0	1	1	1	1	3	4	6
	500	0	0	0	1	1	1	1	3	4	5
	600	0	0	0	0	1	1	1	2	3	4
	700	0	0	0	0	0	1	1	1	3	3
	750	0	0	0	0	0	0	1	1	2	3
	800	0	0	0	0	0	0	1	1	2	3
	900	0	0	0	0	0	0	1	1	2	3
	1000	0	0	0	0	1	1	1	1	1	3
	1250	0	0	0	0	1	0	1	1	1	1
	1500	0	0	0	0	1	0	1	1	1	1
	1750	0	0	0	0	1	0	0	1	1	1
	2000	0	0	0	0	0	0	0	1	1	1

Type	Size										
TW, THHW, THW, THW-2	14	365	280	215	140	93	57	44	25	15	9
	12	280	215	165	108	71	43	33	19	12	7
	10	209	160	123	80	53	32	25	14	9	5
	8	116	89	68	44	29	18	14	8	5	3
RHH,[a] RHW,[a] RHW-2[a]	14	243	186	143	93	62	38	29	16	10	6
	12	195	149	115	75	50	30	23	13	8	5
	10	152	117	89	58	39	23	18	10	6	3
	8	91	70	53	35	23	14	11	6	4	3
RHH,[a] RHW,[a] RHW-2,[a] TW, THW, THHW, THW-2	6	70	53	41	27	18	11	8	5	3	1
	4	52	40	30	20	13	8	6	3	1	1
	3	44	34	26	17	11	7	5	3	1	1
	2	38	29	22	14	9	6	4	2	1	1
	1	26	20	15	10	7	4	3	1	1	1
	1/0	23	17	13	8	6	3	2	1	1	0
	2/0	19	15	11	7	5	3	2	1	1	0
	3/0	16	12	9	6	4	2	1	1	1	0
	4/0	13	10	8	5	3	1	1	1	0	0
	250	11	8	6	4	3	3	2	1	1	0
	300	9	7	5	3	2	2	1	1	0	0
	350	8	6	5	3	1	1	1	0	0	0

Continued

Exhibit VII.51 Maximum Number of Same Size Conductors or Fixture Wires in Liquidtight Flexible Metal Conduit (LFMC) *Continued*

Type	Conductor Size (AWG/kcmil)	Conductors — Metric Designator (Trade Size)									
		16 (1/2)	21 (3/4)	27 (1)	35 (1-1/4)	41 (1-1/2)	53 (2)	63 (2-1/2)	78 (3)	91 (3-1/2)	103 (4)
RHH,[a] RHW,[a] RHW-2,[a]	400	0	0	0	1	1	1	3	4	6	7
	500	0	0	0	1	1	1	2	3	5	6
TW, THW, THHW, THW-2	600	0	0	0	0	1	1	1	3	4	5
	700	0	0	0	0	1	1	1	2	3	4
	750	0	0	0	0	1	1	1	2	3	4
	800	0	0	0	0	1	1	1	2	3	4
	900	0	0	0	0	0	1	1	1	3	3
	1000	0	0	0	0	0	1	1	1	2	3
	1250	0	0	0	0	0	0	1	1	1	2
	1500	0	0	0	0	0	0	1	1	1	2
	1750	0	0	0	0	0	0	1	1	1	1
	2000	0	0	0	0	0	0	1	1	1	1

THHN, THWN, THWN-2										
14	523	401	308	201	133	81	63	36	22	13
12	381	292	225	146	97	59	46	26	16	9
10	240	184	141	92	61	37	29	16	10	6
8	138	106	81	53	35	21	16	9	6	3
6	100	76	59	38	25	15	12	7	4	2
4	61	47	36	23	15	9	7	4	2	1
3	52	40	30	20	13	8	6	3	1	1
2	44	33	26	17	11	7	5	3	1	1
1	32	25	19	12	8	5	4	1	1	1
1/0	27	21	16	10	7	4	3	1	1	1
2/0	23	17	13	8	6	3	2	1	1	0
3/0	19	14	11	7	5	3	1	1	1	0
4/0	15	12	9	6	4	2	1	1	1	0
250	12	10	7	5	3	3	1	1	0	0
300	11	8	6	4	3	3	1	1	0	0
350	9	7	5	3	2	2	1	0	0	0
400	8	6	5	3	1	1	1	0	0	0
500	7	5	4	2	1	1	1	0	0	0
600	6	4	3	1	1	1	1	0	0	0
700	5	4	3	1	1	1	1	0	0	0

Continued

Exhibit VII.51 Maximum Number of Same Size Conductors or Fixture Wires in Liquidtight Flexible Metal Conduit (LFMC) *Continued*

Type	Conductor Size (AWG/kcmil)	Metric Designator (Trade Size)									
		16 (1/2)	21 (3/4)	27 (1)	35 (1-1/4)	41 (1-1/2)	53 (2)	63 (2-1/2)	78 (3)	91 (3-1/2)	103 (4)
THHN, THWN, THWN-2	750	0	0	0	0	1	1	1	3	3	5
	800	0	0	0	0	1	1	1	2	3	4
	900	0	0	0	0	1	1	1	2	3	4
	1000	0	0	0	0	0	1	1	1	3	3
FEP, FEPB, PFA, PFAH, TFE	14	12	21	35	61	79	129	195	299	389	507
	12	9	15	25	44	57	94	142	218	284	370
	10	6	11	18	32	41	68	102	156	203	266
	8	3	6	10	18	23	39	58	89	117	152
	6	2	4	7	13	17	27	41	64	83	108
PFA, PFAH, TFE	4	1	3	5	9	12	19	29	44	58	75
	3	1	2	4	7	10	16	24	37	48	63
	2	1	1	3	6	8	13	20	30	40	52
PFAH, TFE	1	1	1	2	4	5	9	14	21	28	36

PFA, PFAH, TFE, Z	1/0	1	1	1	3	4	7	11	18	23	30
	2/0	1	1	1	3	4	6	9	14	19	25
	3/0	0	1	1	2	3	5	8	12	16	20
	4/0	0	1	1	1	2	4	6	10	13	17
Z	14	20	26	42	73	95	156	235	360	469	611
	12	14	18	30	52	67	111	167	255	332	434
	10	8	11	18	32	41	68	102	156	203	266
	8	5	7	11	20	26	43	64	99	129	168
	6	4	5	8	14	18	30	45	69	90	118
	4	2	3	5	9	12	20	31	48	62	81
	3	2	2	4	7	9	15	23	35	45	59
	2	1	1	3	6	7	12	19	29	38	49
	1	1	1	2	5	6	10	15	23	30	40
XHH, XHHW, XHHW-2, ZW	14	9	15	25	44	57	93	140	215	280	365
	12	7	12	19	33	43	71	108	165	215	280
	10	5	9	14	25	32	53	80	123	160	209
	8	3	5	8	14	18	29	44	68	89	116
	6	1	3	6	10	13	22	33	50	66	86

Continued

Exhibit VII.51 Maximum Number of Same Size Conductors or Fixture Wires in Liquidtight Flexible Metal Conduit (LFMC) *Continued*

Conductors

Type	Conductor Size (AWG/kcmil)	16 (1/2)	21 (3/4)	27 (1)	35 (1-1/4)	41 (1-1/2)	53 (2)	63 (2-1/2)	78 (3)	91 (3-1/2)	103 (4)
					Metric Designator (Trade Size)						
XHH, XHHW, XHHW-2, ZW	4	1	2	4	7	9	16	24	36	48	62
	3	1	1	3	6	8	13	20	31	40	52
	2	1	1	3	5	7	11	17	26	34	44
	1	1	1	1	4	5	8	12	19	25	33
XHH, XHHW, XHHW-2	1/0	1	1	1	3	4	7	10	16	21	28
	2/0	0	1	1	2	3	6	9	13	17	23
	3/0	0	1	1	1	3	5	7	11	14	19
	4/0	0	0	1	1	2	4	6	9	12	16
	250	0	0	1	1	1	3	5	7	10	13
	300	0	0	1	1	1	3	4	6	8	11
	350	0	0	1	1	1	2	3	5	7	10
	400	0	0	0	1	1	1	3	5	6	8
	500	0	0	0	1	1	1	2	4	5	7

Conductor Size[a]								
600	0	0	0	1	1	1	3	4	6
700	0	0	0	1	1	1	3	4	5
750	0	0	0	1	1	1	3	3	5
800	0	0	0	1	1	1	2	3	4
900	0	0	0	1	1	1	2	3	4
1000	0	0	0	0	1	1	1	3	3
1250	0	0	0	0	1	1	1	1	3
1500	0	0	0	0	0	1	1	1	2
1750	0	0	0	0	0	1	1	1	2
2000	0	0	0	0	0	1	1	1	2

[a]Types RHH, RHW, and RHW-2 without outer covering.

Fixture Wires

Type	Conductor Size (AWG/kcmil)	Metric Designator (Trade Size)					
		16 (1/2)	21 (3/4)	27 (1)	35 (1-1/4)	41 (1-1/2)	53 (2)
FFH-2, RFH-2, RFHH-3	18	8	15	24	42	54	89
	16	7	12	20	35	46	75

Continued

Exhibit VII.51 Maximum Number of Same Size Conductors or Fixture Wires in Liquidtight Flexible Metal Conduit (LFMC) *Continued*

Fixture Wires

Type	Conductor Size (AWG/kcmil)	Metric Designator (Trade Size)						
		16 (1/2)	21 (3/4)	27 (1)	35 (1-1/4)	41 (1-1/2)	53 (2)
SF-2, SFF-2	18	11	19	30	53	69	113
	16	9	15	25	44	57	93
	14	7	12	20	35	46	75
SF-1, SFF-1	18	19	33	53	94	122	199
RFH-1, RFHH-2, TF, TFF, XF, XFF	18	14	24	39	69	90	147
RFHH-2, TF, TFF, XF, XFF	16	11	20	32	56	72	119
XF, XFF	14	9	15	25	44	57	93
TFN, TFFN	18	23	39	63	111	144	236
	16	17	30	48	85	110	180
PF, PFF, PGF, PGFF, PAF, PTF, PTFF, PAFF	18	21	37	60	105	136	223
	16	16	29	46	81	105	173
	14	12	21	35	61	79	129

Type	Size						
HF, HFF, ZF, ZFF, ZHF	18	28	48	77	136	176	288
	16	20	35	57	100	129	212
	14	15	26	42	73	95	156
KF-2, KFF-2	18	40	70	112	197	255	418
	16	28	49	79	139	180	295
	14	19	34	54	95	123	202
	12	13	23	37	65	85	139
	10	9	15	25	44	57	93
KF-1, KFF-1	18	48	83	134	235	304	499
	16	34	58	94	165	214	350
	14	23	39	63	111	144	236
	12	15	26	42	73	95	156
	10	10	17	27	48	62	102
XF, XFF	12	5	8	13	23	30	50
	10	3	6	10	18	23	39

Note: This table is for concentric stranded Conductors only. For compact stranded Conductors, Exhibit VII.52 should be used.
Source: NFPA 70, *National Electrical Code®*, NFPA, Quincy, MA, 2002 edition, Table C7.

Exhibit VII.52 Maximum Number of Same Size Compact Conductors in Liquidtight Flexible Metal Conduit (LFMC)

Compact Conductors

Type	Conductor Size (AWG/kcmil)	Metric Designator (Trade Size)										
		12 (3/8)	16 (1/2)	21 (3/4)	27 (1)	35 (1-1/4)	41 (1-1/2)	53 (2)	63 (2-1/2)	78 (3)	91 (3-1/2)	103 (4)
THW, THW-2, THHW	8	1	2	4	7	12	15	25	38	58	76	99
	6	1	1	3	5	9	12	19	29	45	59	77
	4	1	1	2	4	7	9	14	22	34	44	57
	2	1	1	1	3	5	6	11	16	25	32	42
	1	0	1	1	1	3	4	7	11	17	23	30
	1/0	0	1	1	1	3	4	6	10	15	20	26
	2/0	0	0	1	1	2	3	5	8	13	16	21
	3/0	0	0	1	1	1	3	4	7	11	14	18
	4/0	0	0	0	1	1	2	4	6	9	12	15
	250	0	0	0	1	1	1	3	4	7	9	12
	300	0	0	0	1	1	1	2	4	6	8	10
	350	0	0	0	0	1	1	2	3	5	7	9
	400	0	0	0	0	1	1	1	3	5	6	8
	500	0	0	0	0	1	1	1	3	4	5	7

Conductor Size											
600	0	0	0	0	1	1	1	1	3	4	6
700	0	0	0	0	1	1	1	1	3	4	5
750	0	0	0	0	0	1	1	1	3	3	5
1000	0	0	0	0	0	1	1	1	1	3	4
THHN, THWN, THWN-2											
8	—	—	—	—	—	—	—	—	—	—	—
6	1	2	4	7	13	17	28	43	66	86	112
4	1	1	3	4	8	11	17	26	41	53	69
2	1	1	1	3	6	7	12	19	29	38	50
1	0	1	1	2	4	6	9	14	22	28	37
1/0	0	1	1	1	4	5	8	12	19	24	32
2/0	0	1	1	1	3	4	6	10	15	20	26
3/0	0	0	1	1	2	3	5	8	13	17	22
4/0	0	0	1	1	1	3	4	7	10	14	18
250	0	0	1	1	1	1	3	5	8	11	14
300	0	0	0	1	1	1	3	4	7	9	12
350	0	0	0	1	1	1	2	4	6	8	11
400	0	0	0	1	1	1	2	3	5	7	9
500	0	0	0	0	1	1	1	3	5	6	8
600	0	0	0	0	1	1	1	2	4	5	6
700	0	0	0	0	1	1	1	1	3	4	6

Continued

Exhibit VII.52 Maximum Number of Same Size Compact Conductors in Liquidtight Flexible Metal Conduit (LFMC) *Continued*

Compact Conductors

Type	Conductor Size (AWG/kcmil)	Metric Designator (Trade Size)										
		12 (3/8)	16 (1/2)	21 (3/4)	27 (1)	35 (1-1/4)	41 (1-1/2)	53 (2)	63 (2-1/2)	78 (3)	91 (3-1/2)	103 (4)
THHN, THWN, THWN-2	750	0	0	0	0	0	1	1	1	3	4	5
	1000	0	0	0	0	0	1	1	1	2	3	4
XHHW, XHHW-2	8	1	3	5	9	15	20	33	49	76	98	129
	6	1	2	4	6	11	15	24	37	56	73	95
	4	1	1	3	4	8	11	17	26	41	53	69
	2	1	1	1	3	6	7	12	19	29	38	50
	1	0	1	1	2	4	6	9	14	22	28	37
	1/0	0	1	1	1	4	5	8	12	19	24	32
	2/0	0	1	1	1	3	4	7	10	16	20	27
	3/0	0	0	1	1	2	3	5	8	13	17	22
	4/0	0	0	1	1	1	3	4	7	11	14	18

250	0	0	1	1	1	3	5	8	11	15
300	0	0	0	1	1	3	5	7	9	12
350	0	0	0	1	1	3	4	6	8	11
400	0	0	0	1	1	2	4	6	7	10
500	0	0	0	0	1	1	3	5	6	8
600	0	0	0	0	1	1	2	4	5	6
700	0	0	0	0	1	1	1	3	4	5
750	0	0	0	1	1	1	1	3	4	5
1000	0	0	0	0	0	1	1	2	3	4

Note: *Compact stranding* is defined as the result of a manufacturing process where the standard conductor is compressed to the extent that the interstices (voids between strand wires) are virtually eliminated.

Source: NFPA 70, *National Electrical Code®*, NFPA, Quincy, MA, 2002 edition, Table C7(A).

Exhibit VII.53 Maximum Number of Same Size Conductors or Fixture Wires in Rigid Metal Conduit (RMC)

| | Conductor Size (AWG/kcmil) | Conductors — Metric Designator (Trade Size) | | | | | | | | | | | |
Type		16 (1/2)	21 (3/4)	27 (1)	35 (1-1/4)	41 (1-1/2)	53 (2)	63 (2-1/2)	78 (3)	91 (3-1/2)	103 (4)	129 (5)	155 (6)
RHH, RHW, RHW-2	14	4	7	12	21	28	46	66	102	136	176	276	398
	12	3	6	10	17	23	38	55	85	113	146	229	330
	10	3	5	8	14	19	31	44	68	91	118	185	267
	8	1	2	4	7	10	16	23	36	48	61	97	139
	6	1	1	3	6	8	13	18	29	38	49	77	112
	4	1	1	2	4	6	10	14	22	30	38	60	87
	3	1	1	2	4	5	9	12	19	26	34	53	76
	2	1	1	1	3	4	7	11	17	23	29	46	66
	1	0	1	1	1	3	5	7	11	15	19	30	44
	1/0	0	1	1	1	2	4	6	10	13	17	26	38
	2/0	0	1	1	1	2	4	5	8	11	14	23	33
	3/0	0	0	1	1	1	3	4	7	10	12	20	28
	4/0	0	0	1	1	1	3	4	6	8	11	17	24
	250	0	0	0	1	1	1	3	4	6	8	13	18

300	16	11	7	5	4	2	1	1	1	0	0	0
350	15	10	6	5	4	2	1	1	1	0	0	0
400	13	9	6	4	3	1	1	1	1	0	0	0
500	11	8	5	4	3	1	1	1	1	0	0	0
600	9	6	4	3	2	1	1	1	0	0	0	0
700	8	6	4	3	1	1	1	1	0	0	0	0
750	8	5	3	3	1	1	1	0	0	0	0	0
800	7	5	3	2	1	1	1	0	0	0	0	0
900	7	5	3	2	1	1	1	0	0	0	0	0
1000	6	4	3	1	1	1	0	0	0	0	0	0
1250	5	3	1	1	1	1	0	0	0	0	0	0
1500	4	3	1	1	1	1	0	0	0	0	0	0
1750	4	2	1	1	1	1	0	0	0	0	0	0
2000	3	2	1	1	1	0	0	0	0	0	0	0
TW, THHW, THW, THW-2												
14	839	581	370	288	216	140	98	59	44	25	15	9
12	644	446	284	221	165	107	75	45	33	19	12	7
10	480	332	212	164	123	80	56	34	25	14	9	5
8	267	185	118	91	68	44	31	19	14	8	5	3
RHH,[a] RHW,[a] RHW-2[a]												
14	558	387	246	191	143	93	65	39	29	17	10	6

Continued

Exhibit VII.53 Maximum Number of Same Size Conductors or Fixture Wires in Rigid Metal Conduit (RMC) *Continued*

	Conductor Size (AWG/kcmil)	Conductors — Metric Designator (Trade Size)											
Type		16 (1/2)	21 (3/4)	27 (1)	35 (1-1/4)	41 (1-1/2)	53 (2)	63 (2-1/2)	78 (3)	91 (3-1/2)	103 (4)	129 (5)	155 (6)
RHH,[a] RHW,[a]	12	5	8	13	23	32	52	75	115	154	198	311	448
RHW-2[a]	10	3	6	10	18	25	41	58	90	120	154	242	350
RHH, RHW,[a] RHW-2[a]	8	1	4	6	11	15	24	35	54	72	92	145	209
RHH,[a] RHW,[a]	6	1	3	5	8	11	18	27	41	55	71	111	160
RHW-2,[a]	4	1	1	3	6	8	14	20	31	41	53	83	120
TW, THW,	3	1	1	3	5	7	12	17	26	35	45	71	103
THHW,	2	1	1	2	4	6	10	14	22	30	38	60	87
THW-2	1	1	1	1	3	4	7	10	15	21	27	42	61
	1/0	0	1	1	2	3	6	8	13	18	23	36	52
	2/0	0	1	1	2	3	5	7	11	15	19	31	44
	3/0	0	1	1	1	2	4	6	9	13	16	26	37
	4/0	0	0	1	1	1	3	5	8	10	14	21	31

250	0	0	1	1	1	3	4	6	8	11	17	25
300	0	0	1	1	1	2	3	5	7	9	15	22
350	0	0	0	1	1	1	3	5	6	8	13	19
400	0	0	0	1	1	1	3	4	6	7	12	17
500	0	0	0	1	1	1	2	3	5	6	10	14
600	0	0	0	1	1	1	1	3	4	5	8	12
700	0	0	0	0	1	1	1	2	3	4	7	10
750	0	0	0	0	1	1	1	2	3	4	7	10
800	0	0	0	0	1	1	1	2	3	4	6	9
900	0	0	0	0	1	1	1	1	3	4	6	8
1000	0	0	0	0	0	1	1	1	2	3	5	8
1250	0	0	0	0	0	1	1	1	1	2	4	6
1500	0	0	0	0	0	1	1	1	1	2	3	5
1750	0	0	0	0	0	0	1	1	1	1	3	4
2000	0	0	0	0	0	0	1	1	1	1	3	4
14	13	22	36	63	85	140	200	309	412	531	833	1202
12	9	16	26	46	62	102	146	225	301	387	608	877
10	6	10	17	29	39	64	92	142	189	244	383	552
8	3	6	9	16	22	37	53	82	109	140	221	318
6	2	4	7	12	16	27	38	59	79	101	159	230

THHN,
THWN,
THWN-2

Continued

Exhibit VII.53 Maximum Number of Same Size Conductors or Fixture Wires in Rigid Metal Conduit (RMC) *Continued*

Type	Conductor Size (AWG/kcmil)	Conductors / Metric Designator (Trade Size)											
		16 (1/2)	21 (3/4)	27 (1)	35 (1-1/4)	41 (1-1/2)	53 (2)	63 (2-1/2)	78 (3)	91 (3-1/2)	103 (4)	129 (5)	155 (6)
THHN, THWN, THWN-2	4	1	2	4	7	10	16	23	36	48	62	98	141
	3	1	1	3	6	8	14	20	31	41	53	83	120
	2	1	1	3	5	7	11	17	26	34	44	70	100
	1	1	1	1	4	5	8	12	19	25	33	51	74
	1/0	1	1	1	3	4	7	10	16	21	27	43	63
	2/0	0	1	1	2	3	6	8	13	18	23	36	52
	3/0	0	1	1	1	3	5	7	11	15	19	30	43
	4/0	0	1	1	1	2	4	6	9	12	16	25	36
	250	0	0	1	1	1	3	5	7	10	13	20	29
	300	0	0	1	1	1	3	4	6	8	11	17	25
	350	0	0	1	1	1	2	3	5	7	10	15	22
	400	0	0	1	1	1	2	3	5	7	8	13	20
	500	0	0	0	1	1	1	2	4	5	7	11	16

Continued

Type	Size												
	600	0	0	0	1	1	1	1	3	4	6	9	13
	700	0	0	0	1	1	1	1	3	4	5	8	11
	750	0	0	0	0	1	1	1	3	4	5	7	11
	800	0	0	0	0	1	1	1	2	3	4	7	10
	900	0	0	0	0	1	1	1	2	3	4	6	9
	1000	0	0	0	0	1	1	1	1	3	4	6	8
FEP, FEPB, PFA, PFAH, TFE	14	12	22	35	61	83	136	194	300	400	515	808	1166
	12	9	16	26	44	60	99	142	219	292	376	590	851
	10	6	11	18	32	43	71	102	157	209	269	423	610
	8	3	6	10	18	25	41	58	90	120	154	242	350
	6	2	4	7	13	17	29	41	64	85	110	172	249
	4	1	3	5	9	12	20	29	44	59	77	120	174
	3	1	2	4	7	10	17	24	37	50	64	100	145
	2	1	1	3	6	8	14	20	31	41	53	83	120
PFA, PFAH, TFE	1	1	1	2	4	6	9	14	21	28	37	57	83
PFA, PFAH, TFE, Z	1/0	1	1	1	3	5	8	11	18	24	30	48	69
	2/0	1	1	1	3	4	6	9	14	19	25	40	57

Exhibit VII.53 Maximum Number of Same Size Conductors or Fixture Wires in Rigid Metal Conduit (RMC) *Continued*

Conductors

Type	Conductor Size (AWG/kcmil)	Metric Designator (Trade Size)											
		16 (1/2)	21 (3/4)	27 (1)	35 (1-1/4)	41 (1-1/2)	53 (2)	63 (2-1/2)	78 (3)	91 (3-1/2)	103 (4)	129 (5)	155 (6)
PFA, PFAH, TFE, Z	3/0	0	1	1	2	3	5	6	12	16	21	33	47
	4/0	0	1	1	1	2	4	6	10	13	17	27	39
Z	14	15	26	42	73	100	164	234	361	482	621	974	1405
	12	10	18	30	52	71	116	166	256	342	440	691	997
	10	6	11	18	32	43	71	102	157	209	269	423	610
	8	4	7	11	20	27	45	64	99	132	170	267	386
	6	3	5	8	14	19	31	45	69	93	120	188	271
	4	1	3	5	9	13	22	31	48	64	82	129	186
	3	1	2	4	7	9	16	22	35	47	60	94	136
	2	1	1	3	6	8	13	19	29	39	50	78	113
	1	1	1	2	5	6	10	15	23	31	40	63	92
XHH, XHHW, XHHW-2, ZW	14	9	15	25	44	59	98	140	216	288	370	581	839
	12	7	12	19	33	45	75	107	165	221	284	446	644
	10	5	9	14	25	34	56	80	123	164	212	332	480
	8	3	5	8	14	19	31	44	68	91	118	185	267

		1	3	6	10	14	23	33	51	68	87	137	197
	6	1	3	6	10	14	23	33	51	68	87	137	197
	4	1	2	4	7	10	16	24	37	49	63	99	143
	3	1	1	3	6	8	14	20	31	41	53	84	121
	2	1	1	3	5	7	12	17	26	35	45	70	101
XHH, XHHW, XHHW-2	1	1	1	1	4	5	9	12	19	26	33	52	76
	1/0	1	1	1	3	4	7	10	16	22	28	44	64
	2/0	0	1	1	2	3	6	9	13	18	23	37	53
	3/0	0	1	1	1	3	5	7	11	15	19	30	44
	4/0	0	1	1	1	2	4	6	10	12	16	25	36
	250	0	0	1	1	1	3	5	7	10	13	20	30
	300	0	0	1	1	1	3	4	6	9	11	18	25
	350	0	0	1	1	1	2	4	6	7	10	15	22
	400	0	0	1	1	1	2	3	5	7	9	14	20
	500	0	0	0	1	1	1	2	4	5	7	11	16
	600	0	0	0	1	1	1	1	3	4	6	9	13
	700	0	0	0	1	1	1	1	3	4	5	8	11
	750	0	0	0	0	1	1	1	2	3	5	7	11
	800	0	0	0	0	1	1	1	2	3	4	7	10
	900	0	0	0	0	1	1	1	2	3	4	6	9

Continued

Exhibit VII.53 Maximum Number of Same Size Conductors or Fixture Wires in Rigid Metal Conduit (RMC) *Continued*

Conductors

Type	Conductor Size (AWG/kcmil)	Metric Designator (Trade Size)											
		16 (1/2)	21 (3/4)	27 (1)	35 (1-1/4)	41 (1-1/2)	53 (2)	63 (2-1/2)	78 (3)	91 (3-1/2)	103 (4)	129 (5)	155 (6)
XHH, XHHW, XHHW-2	1000	0	0	0	0	0	0	1	1	3	4	6	8
	1250	0	0	0	0	0	0	1	1	2	3	4	6
	1500	0	0	0	0	0	1	1	1	1	2	4	5
	1750	0	0	0	0	0	1	1	1	1	1	3	5
	2000	0	0	0	0	0	1	1	1	1	1	3	4

Fixture Wires

Type	Conductor Size (AWG/kcmil)	Metric Designator (Trade Size)					
		16 (1/2)	21 (3/4)	27 (1)	35 (1-1/4)	41 (1-1/2)	53 (2)
FFH-2, RFH-2, RFHH-3	18	8	15	24	42	57	94
	16	7	12	20	35	48	79

SF-2, SFF-2	18	11	19	31	53	72	118
	16	9	15	25	44	59	98
	14	7	12	20	35	48	79
SF-1, SFF-1	18	19	33	54	94	127	209
RFH-1, RFHH-2, TF, TFF, XF, XFF	18	14	25	40	69	94	155
RFHH-2, TF, TFF, XF, XFF	16	11	20	32	56	76	125
XF, XFF	14	9	15	25	44	59	98
TFN, TFFN	18	23	40	64	111	150	248
	16	17	30	49	84	115	189
PF, PFF, PGF, PGFF, PAF, PTF, PTFF, PAFF	18	21	38	61	105	143	235
	16	16	29	47	81	110	181
	14	12	22	35	61	83	136
HF, HFF, ZF, ZFF, ZHF	18	28	48	79	135	184	303
	16	20	36	58	100	136	223
	14	15	26	42	73	100	164
KF-2, KFF-2	18	40	71	114	197	267	439
	16	28	50	80	138	188	310
	14	19	34	55	95	129	213

Continued

Exhibit VII.53 Maximum Number of Same Size Conductors or Fixture Wires in Rigid Metal Conduit (RMC) *Continued*

Fixture Wires

Type	Conductor Size (AWG/kcmil)	Metric Designator (Trade Size)						
		16 (1/2)	21 (3/4)	27 (1)	35 (1-1/4)	41 (1-1/2)	53 (2)
KF-2, KFF-2	12	13	23	38	65	89	146
	10	9	15	25	44	59	98
KF-1, KFF-1	18	48	84	136	235	318	524
	16	34	59	96	165	224	368
	14	23	40	64	111	150	248
	12	15	26	42	73	100	164
	10	10	17	28	48	65	107
XF, XFF	12	5	8	13	23	32	52
	10	3	6	10	18	25	41

Note: This table is for concentric stranded Conductors only. For compact stranded Conductors, Exhibit VII.54 should be used.

[a]Types RHH, RHW, and RHW-2 without outer covering.

Source: NFPA 70, *National Electrical Code®*, NFPA, Quincy, MA, 2002 edition, Table C8.

Exhibit VII.54 Maximum Number of Same Size Compact Conductors in Rigid Metal Conduit (RMC)

Conductor Size (AWG/kcmil)	Compact Conductors											
	Metric Designator (Trade Size)											
	16 (1/2)	21 (3/4)	27 (1)	35 (1-1/4)	41 (1-1/2)	53 (2)	63 (2-1/2)	78 (3)	91 (3-1/2)	103 (4)	129 (5)	155 (6)
Type Letters												
THW, THW-2, THHW												
8	2	4	7	12	16	26	38	59	78	101	158	228
6	1	3	5	9	12	20	29	45	60	78	122	176
4	1	2	4	7	9	15	22	34	45	58	91	132
2	1	1	3	5	7	11	16	25	33	43	67	97
1	1	1	1	3	5	8	11	17	23	30	47	68
1/0	0	1	1	3	4	6	10	15	20	26	41	59
2/0	0	1	1	2	3	6	8	13	17	22	34	50
3/0	0	1	1	1	3	5	7	11	14	19	29	42
4/0	0	1	1	1	2	4	6	9	12	15	24	35
250	0	0	1	1	1	3	4	7	9	12	19	28
300	0	0	1	1	1	3	4	6	8	11	17	24
350	0	0	0	1	1	2	3	5	7	9	15	22

Continued

Exhibit VII.54 Maximum Number of Same Size Compact Conductors in Rigid Metal Conduit (RMC) *Continued*

Compact Conductors

Type Letters	Conductor Size (AWG/kcmil)	Metric Designator (Trade Size)											
		16 (1/2)	21 (3/4)	27 (1)	35 (1-1/4)	41 (1-1/2)	53 (2)	63 (2-1/2)	78 (3)	91 (3-1/2)	103 (4)	129 (5)	155 (6)
THW, THW-2, THHW	400	0	0	1	1	1	1	3	5	7	8	13	20
	500	0	0	0	1	1	1	3	4	5	7	11	17
	600	0	0	0	1	1	1	1	3	4	6	9	13
	700	0	0	0	1	1	1	1	3	4	5	8	12
	750	0	0	0	0	1	1	1	3	4	5	7	11
	1000	0	0	0	0	1	1	1	1	3	4	6	9
THHN, THWN, THWN-2	8	—	—	—	—	—	—	—	—	—	—	—	—
	6	2	5	8	13	18	30	43	66	88	114	179	258
	4	1	3	5	8	11	18	26	41	55	70	110	159
	2	1	1	3	6	8	13	19	29	39	50	79	114
	1	1	1	2	4	6	10	14	22	29	38	60	86
	1/0	1	1	1	3	5	8	12	19	25	32	51	73
	2/0	1	1	1	3	4	7	10	15	21	26	42	60
	3/0	0	1	1	2	3	6	8	13	17	22	35	51
	4/0	0	1	1	1	3	5	7	10	14	18	29	42

XHHW, XHHW-2

Size												
250	33	23	14	11	8	5	4	2	1	1	0	0
300	28	20	12	10	7	4	3	1	1	1	0	0
350	25	17	11	8	6	4	3	1	1	1	0	0
400	22	15	10	7	5	3	2	1	1	1	0	0
500	19	13	8	6	5	3	1	1	1	0	0	0
600	15	10	6	5	4	2	1	1	1	0	0	0
700	13	9	6	4	3	1	1	1	1	0	0	0
750	13	9	5	4	3	1	1	1	1	0	0	0
1000	9	6	4	3	2	1	1	1	0	0	0	0
8	296	205	130	101	76	49	34	21	15	9	5	3
6	220	152	97	75	56	36	25	15	11	6	4	2
4	159	110	70	55	41	26	18	11	8	5	3	1
2	114	79	50	39	29	19	13	8	6	3	1	1
1		60	38	29	22	14	10	6	4	2	1	1
1/0	73	51	32	25	19	12	8	5	4	1	1	1
2/0	62	43	27	21	16	10	7	4	3	1	1	1
3/0	51	35	22	17	13	8	6	3	2	1	1	0
4/0	42	29	19	14	11	7	5	3	1	1	1	0
250	34	23	15	11	8	5	4	2	1	1	1	0
300	29	20	13	10	7	5	3	1	1	1	0	0

Continued

Exhibit VII.54 Maximum Number of Same Size Compact Conductors in Rigid Metal Conduit (RMC) *Continued*

Compact Conductors

Type Letters	Conductor Size (AWG/kcmil)	Metric Designator (Trade Size)											
		16 (1/2)	21 (3/4)	27 (1)	35 (1-1/4)	41 (1-1/2)	53 (2)	63 (2-1/2)	78 (3)	91 (3-1/2)	103 (4)	129 (5)	155 (6)
XHHW, XHHW-2	350	0	0	1	1	1	3	4	6	9	11	18	25
	400	0	0	1	1	1	2	4	6	8	10	16	23
	500	0	0	0	1	1	1	3	5	6	8	13	19
	600	0	0	0	1	1	1	2	4	5	7	10	15
	700	0	0	0	1	1	1	1	3	4	6	9	13
	750	0	0	0	1	1	1	1	3	4	5	8	12
	1000	0	0	0	0	1	1	1	2	3	4	7	10

Note: Compact stranding is defined as the result of a manufacturing process where the standard conductor is compressed to the extent that the interstices (voids between strand wires) are virtually eliminated.

Source: NFPA 70, National Electrical Code®, NFPA, Quincy, MA, 2002 edition, Table C8(A).

Exhibit VII.55 Maximum Number of Same Size Conductors or Fixture Wires in Rigid PVC Conduit, Schedule 80

Type	Conductor Size (AWG/kcmil)	Conductors — Metric Designator (Trade Size)											
		16 (1/2)	21 (3/4)	27 (1)	35 (1-1/4)	41 (1-1/2)	53 (2)	63 (2-1/2)	78 (3)	91 (3-1/2)	103 (4)	129 (5)	155 (6)
RHH, RHW, RHW-2	14	3	5	9	17	23	39	56	88	118	153	243	349
	12	2	4	7	14	19	32	46	73	98	127	202	290
	10	1	3	6	11	15	26	37	59	79	103	163	234
	8	1	1	3	6	8	13	19	31	41	54	85	122
	6	1	1	2	4	6	11	16	24	33	43	68	98
	4	1	1	1	3	5	8	12	19	26	33	53	77
	3	0	1	1	3	4	7	11	17	23	29	47	67
	2	0	1	1	3	4	6	9	14	20	25	41	58
	1	0	1	1	1	2	4	6	9	13	17	27	38
	1/0	0	0	1	1	3	3	5	8	11	15	23	33
	2/0	0	0	1	1	3	3	4	7	10	13	20	29
	3/0	0	0	0	1	1	3	4	6	8	11	17	25
	4/0	0	0	0	1	1	2	3	5	7	9	15	21

Continued

Exhibit VII.55 Maximum Number of Same Size Conductors or Fixture Wires in Rigid PVC Conduit, Schedule 80 *Continued*

Conductors

Type	Conductor Size (AWG/kcmil)	Metric Designator (Trade Size)											
		16 (1/2)	21 (3/4)	27 (1)	35 (1-1/4)	41 (1-1/2)	53 (2)	63 (2-1/2)	78 (3)	91 (3-1/2)	103 (4)	129 (5)	155 (6)
RHH, RHW, RHW-2	250	0	0	0	1	1	1	2	4	5	7	11	16
	300	0	0	0	1	1	1	2	3	5	6	10	14
	350	0	0	0	1	1	1	1	3	4	5	9	13
	400	0	0	0	0	1	1	1	3	4	5	8	12
	500	0	0	0	0	1	1	1	2	3	4	7	10
	600	0	0	0	0	0	1	1	1	3	3	6	8
	700	0	0	0	0	0	1	1	1	2	3	5	7
	750	0	0	0	0	1	1	1	1	2	3	5	7
	800	0	0	0	0	0	1	1	1	2	3	4	7
	1000	0	0	0	0	0	0	1	1	1	2	4	5
	1250	0	0	0	0	0	0	1	1	1	1	3	4
	1500	0	0	0	0	0	0	0	1	1	2	2	4
	1750	0	0	0	0	0	0	0	1	1	1	2	3
	2000	0	0	0	0	0	0	0	1	1	1	1	3

Type	Size												
TW, THHW, THW, THW-2	14	6	11	20	35	49	82	118	185	250	324	514	736
	12	5	9	15	27	38	63	91	142	192	248	394	565
	10	3	6	11	20	28	47	67	106	143	185	294	421
	8	1	3	6	11	15	26	37	59	79	103	163	234
RHH,[a] RHW,[a] RHW-2[a]	14	4	8	13	23	32	55	79	123	166	215	341	490
	12	3	6	10	19	26	44	63	99	133	173	274	394
	10	2	5	8	15	20	34	49	77	104	135	214	307
	8	1	3	5	9	12	20	29	46	62	81	128	184
RHH,[a] RHW,[a] RHW-2,[a] TW, THHW, THW-2	6	1	1	3	7	9	16	22	35	48	62	98	141
	4	1	1	3	5	7	12	17	26	35	46	73	105
	3	1	1	2	4	6	10	14	22	30	39	63	90
	2	1	1	1	3	5	8	12	19	26	33	53	77
	1	0	1	1	2	3	6	8	13	18	23	37	54
	1/0	0	1	1	1	3	5	7	11	15	20	32	46
	2/0	0	1	1	1	2	4	6	10	13	17	27	39
	3/0	0	0	1	1	1	3	5	8	11	14	23	33
	4/0	0	0	1	1	1	3	4	7	9	12	19	27
	250	0	0	0	1	1	2	3	5	7	9	15	22
	300	0	0	0	1	1	1	3	5	6	8	13	19
	350	0	0	0	1	1	1	2	4	6	7	12	17

Continued

Exhibit VII.55 Maximum Number of Same Size Conductors or Fixture Wires in Rigid PVC Conduit, Schedule 80 *Continued*

Conductors

Type	Conductor Size (AWG/kcmil)	Metric Designator (Trade Size)											
		16 (1/2)	21 (3/4)	27 (1)	35 (1-1/4)	41 (1-1/2)	53 (2)	63 (2-1/2)	78 (3)	91 (3-1/2)	103 (4)	129 (5)	155 (6)
RHH,[a] RHW,[a] RHW-2,[a]	400	0	0	0	1	1	1	2	4	5	7	10	15
	500	0	0	0	1	1	1	1	3	4	5	9	13
TW, THW, THHW, THW-2	600	0	0	0	0	1	1	1	2	3	4	7	10
	700	0	0	0	0	1	1	1	2	3	4	6	9
	750	0	0	0	0	0	1	1	1	3	4	6	8
	800	0	0	0	0	0	1	1	1	3	3	6	8
	900	0	0	0	0	0	1	1	1	2	3	5	7
	1000	0	0	0	0	0	1	1	2	2	3	5	7
	1250	0	0	0	0	0	1	1	1	1	2	4	5
	1500	0	0	0	0	0	0	1	1	1	1	3	4
	1750	0	0	0	0	0	0	1	1	1	1	3	4
	2000	0	0	0	0	0	0	0	1	1	1	2	3

THHN, THWN, THWN-2												
14	1055	736	464	358	265	170	118	70	51	28	17	9
12	770	537	338	261	193	124	86	51	37	20	12	6
10	485	338	213	164	122	78	54	32	23	13	7	4
8	279	195	123	95	70	45	31	18	13	7	4	2
6	202	141	89	68	51	32	22	13	9	5	3	1
4	124	86	54	42	31	20	14	8	6	3	1	1
3	105	73	46	35	26	17	12	7	5	3	1	1
2	88	61	39	30	22	14	10	6	4	2	1	1
1	65	45	29	22	16	10	7	4	3	1	1	0
1/0	55	38	24	18	14	9	6	3	2	1	1	0
2/0	46	32	20	15	11	7	5	3	1	1	1	0
3/0	38	26	17	13	9	6	4	2	1	1	0	0
4/0	31	22	14	10	8	5	3	1	1	1	0	0
250	25	18	11	8	6	4	3	1	1	1	0	0
300	22	15	9	7	5	3	2	1	1	0	0	0
350	19	13	8	6	5	3	1	1	1	0	0	0
400	17	12	7	6	4	3	1	1	1	0	0	0
500	14	10	6	5	3	2	1	1	1	0	0	0
600	12	8	5	4	3	1	1	1	0	0	0	0
700	10	7	4	3	2	1	1	1	0	0	0	0

Continued

Exhibit VII.55 Maximum Number of Same Size Conductors or Fixture Wires in Rigid PVC Conduit, Schedule 80 *Continued*

Type	Conductor Size (AWG/kcmil)	Conductors											
		Metric Designator (Trade Size)											
		16 (1/2)	21 (3/4)	27 (1)	35 (1-1/4)	41 (1-1/2)	53 (2)	63 (2-1/2)	78 (3)	91 (3-1/2)	103 (4)	129 (5)	155 (6)
THHN, THWN, THWN-2	750	0	0	0	0	1	1	1	2	3	4	7	9
	800	0	0	0	0	1	1	1	2	3	4	6	9
	900	0	0	0	0	0	1	1	1	3	3	6	8
	1000	0	0	0	0	0	1	1	1	2	3	5	7
FEP, FEPB, PFA, PFAH, TFE	14	8	16	27	49	68	115	164	257	347	450	714	1024
	12	6	12	20	36	50	84	120	188	253	328	521	747
	10	4	8	14	26	36	60	84	135	182	235	374	536
	8	2	5	8	15	20	34	49	77	104	135	214	307
	6	1	3	6	10	14	24	35	55	74	96	152	218
	4	1	2	4	7	10	17	24	38	52	67	106	153
	3	1	1	3	6	8	14	20	32	43	56	89	127
	2	1	1	3	5	7	12	17	26	35	46	73	105

Type	Size	1	1	1	3	5	8	11	18	25	32	51	73
PFA, PFAH, TFE	1	1	1	1	3	4	7	10	15	20	27	42	61
PFA, PFAH, TFE, Z	1/0	0	1	1	3	3	5	8	12	17	22	35	50
	2/0	0	1	1	2	2	4	6	10	14	18	29	41
	3/0	0	0	1	1	1	4	5	8	11	15	24	34
	4/0	0	0	1	1	1	4	5	8	11	15	24	34
Z	14	10	19	33	59	82	138	198	310	418	542	860	1233
	12	7	14	23	42	58	98	141	220	297	385	610	875
	10	4	8	14	26	36	60	86	135	182	235	374	536
	8	3	5	9	16	22	38	54	85	115	149	236	339
	6	2	4	6	11	16	26	38	60	81	104	166	238
	4	1	2	4	8	11	18	26	41	55	72	114	164
	3	1	2	3	5	8	13	19	30	40	52	83	119
	2	1	1	2	5	6	11	16	25	33	43	69	99
	1	0	1	2	4	5	9	13	20	27	35	56	80
XHH, XHHW, XHHW-2, ZW	14	6	11	20	35	49	82	118	185	250	324	514	736
	12	5	9	15	27	38	63	91	142	192	248	394	565
	10	3	6	11	20	28	47	67	106	143	185	294	421
	8	1	3	6	11	15	26	37	59	79	103	163	234
	6	1	2	4	8	11	19	28	43	59	76	121	173

Continued

Exhibit VII.55 Maximum Number of Same Size Conductors or Fixture Wires in Rigid PVC Conduit, Schedule 80 *Continued*

		Conductors											
		Metric Designator (Trade Size)											
Type	Conductor Size (AWG/kcmil)	16 (1/2)	21 (3/4)	27 (1)	35 (1-1/4)	41 (1-1/2)	53 (2)	63 (2-1/2)	78 (3)	91 (3-1/2)	103 (4)	129 (5)	155 (6)
XHH, XHHW, XHHW-2, ZW	4	1	1	3	6	8	14	20	31	42	55	87	125
	3	1	1	3	5	7	12	17	26	36	47	74	106
	2	1	1	2	4	6	10	14	22	30	39	62	89
	1	0	1	1	3	4	7	10	16	22	29	46	66
XHH, XHHW, XHHW-2	1/0	0	0	1	2	3	6	9	14	19	24	39	56
	2/0	0	0	1	1	3	5	7	11	16	20	32	46
	3/0	0	0	1	1	2	4	6	9	13	17	27	38
	4/0	0	0	1	1	1	3	5	8	11	14	22	32
	250	0	0	1	1	1	3	4	6	9	11	18	26
	300	0	0	0	1	1	2	3	5	7	10	15	22
	350	0	0	0	1	1	1	3	5	6	8	14	20
	400	0	0	0	1	1	1	3	4	6	7	12	17
	500	0	0	0	1	1	1	2	3	5	6	10	14

600	0	0	0	1	1	3	5	8	11
700	0	0	0	1	1	2	4	7	10
750	0	0	0	1	1	2	4	6	9
800	0	0	0	1	1	2	3	6	9
900	0	0	0	1	1	1	—	3	5	8
1000	0	0	0	1	1	2	3	5	7
1250	0	0	0	1	1	1	2	4	6
1500	0	0	0	0	1	1	1	3	5
1750	0	0	0	0	1	1	1	3	4
2000	0	0	0	0	1	1	1	2	4

Fixture Wires

Type	Conductor Size (AWG/kcmil)	Metric Designator (Trade Size)					
		16 (1/2)	21 (3/4)	27 (1)	35 (1-1/4)	41 (1-1/2)	53 (2)
FFH-2, RFH-2, RFHH-3	18	6	11	19	34	47	79
	16	5	9	16	28	39	67

Continued

CONDUIT AND RACEWAY SYSTEMS

Exhibit VII.55 Maximum Number of Same Size Conductors or Fixture Wires in Rigid PVC Conduit, Schedule 80 *Continued*

Fixture Wires

Type	Conductor Size (AWG/kcmil)	16 (1/2)	21 (3/4)	27 (1)	35 (1-1/4)	41 (1-1/2)	53 (2)
SF-2, SFF-2	18	7	14	24	43	59	100
	16	6	11	20	35	49	82
	14	5	9	16	28	39	67
SF-1, SFF-1	18	13	25	42	76	105	177
RFH-1, RFHH-2, TF, TFF, XF, XFF	18	10	18	31	56	77	130
RFHH-2, TF, TFF, XF, XFF	16	8	15	25	45	62	105
XF, XFF	14	6	11	20	35	49	82
TFN, TFFN	18	16	29	50	90	124	209
	16	12	22	38	68	95	159
PF, PFF, PGF, PGFF, PAF, PTF, PTFF, PAFF	18	15	28	47	85	118	198
	16	11	22	36	66	91	153
	14	8	16	27	49	68	115

HF, HFF, ZF, ZFF, ZHF						
18	19	36	61	110	152	255
16	14	27	45	81	112	188
14	10	19	33	59	82	138
KF-2, KFF-2						
18	28	53	88	159	220	371
16	19	37	62	112	155	261
14	13	25	43	77	107	179
12	9	17	29	53	73	123
10	6	11	20	35	49	82
KF-1, KFF-1						
18	33	63	106	190	263	442
16	23	44	74	133	185	310
14	16	29	50	90	124	209
12	10	19	33	59	82	138
10	7	13	21	39	54	90
XF, XFF						
12	3	6	10	19	26	44
10	2	5	8	15	20	34

Note: This table is for concentric stranded Conductors only. For compact stranded Conductors, Exhibit VII.56 should be used.
[a]Types RHH, RHW, and RHW-2 without outer covering.
Source: NFPA 70, *National Electrical Code®*, NFPA, Quincy, MA, 2002 edition, Table C9.

Exhibit VII.56 Maximum Number of Same Size Compact Conductors in Rigid PVC Conduit, Schedule 80

Compact Conductors

Type	Conductor Size (AWG/kcmil)	Metric Designator (Trade Size)											
		16 (1/2)	21 (3/4)	27 (1)	35 (1-1/4)	41 (1-1/2)	53 (2)	63 (2-1/2)	78 (3)	91 (3-1/2)	103 (4)	129 (5)	155 (6)
THW, THW-2, THHW	8	1	3	5	9	13	22	32	50	68	88	140	200
	6	1	2	4	7	10	17	25	39	52	68	108	155
	4	1	1	3	5	7	13	18	29	39	51	81	116
	2	1	1	1	4	5	9	13	21	29	37	60	85
	1	0	1	1	3	4	6	9	15	20	26	42	60
	1/0	0	1	1	2	3	6	8	13	17	23	36	52
	2/0	0	0	1	1	3	5	7	11	15	19	30	44
	3/0	0	0	1	1	2	4	6	9	12	16	26	37
	4/0	0	0	0	1	1	3	5	8	10	13	22	31
	250	0	0	1	1	1	2	4	6	8	11	17	25
	300	0	0	0	1	1	2	3	5	7	9	15	21
	350	0	0	0	1	1	1	3	5	6	8	13	19
	400	0	0	0	1	1	1	3	4	6	7	12	17
	500	0	0	0	1	1	1	2	3	5	6	10	14

	600	0	0	0	0	1	1	1	3	4	5	8	12
	700	0	0	0	0	1	1	1	2	3	4	7	10
	750	0	0	0	0	1	1	1	2	3	4	7	10
	1000	0	0	0	0	0	1	1	1	2	3	5	8
THHN, THWN, THWN-2	8	—	—	—	—	—	—	—	—	—	—	—	—
	6	1	3	6	11	15	25	36	57	77	99	158	226
	4	1	1	3	6	9	15	22	35	47	61	98	140
	2	1	1	2	5	6	11	16	25	34	44	70	100
	1	1	1	1	3	5	8	12	19	25	33	53	75
	1/0	0	1	1	3	4	7	10	16	22	28	45	64
	2/0	0	1	1	2	3	6	8	13	18	23	37	53
	3/0	0	1	1	1	3	5	7	11	15	19	31	44
	4/0	0	0	1	1	2	4	6	9	12	16	25	37
	250	0	0	1	1	1	3	4	7	10	12	20	29
	300	0	0	1	1	1	3	4	6	8	11	17	25
	350	0	0	0	1	1	2	3	5	7	9	15	22
	400	0	0	0	1	1	1	3	5	6	8	13	19
	500	0	0	0	1	1	1	2	4	5	7	11	16

Continued

Exhibit VII.56 Maximum Number of Same Size Compact Conductors in Rigid PVC Conduit, Schedule 80 *Continued*

Compact Conductors

Type	Conductor Size (AWG/kcmil)	Metric Designator (Trade Size)											
		16 (1/2)	21 (3/4)	27 (1)	35 (1-1/4)	41 (1-1/2)	53 (2)	63 (2-1/2)	78 (3)	91 (3-1/2)	103 (4)	129 (5)	155 (6)
THHN, THWN, THWN-2	600	0	0	0	0	1	1	1	3	4	6	9	13
	700	0	0	0	0	1	1	1	3	4	5	8	12
	750	0	0	0	0	0	1	1	3	4	5	8	11
	1000	0	0	0	0	0	1	1	1	3	3	5	8
XHHW, XHHW-2	8	1	4	7	12	17	29	42	65	88	114	181	260
	6	1	3	5	9	13	21	31	48	65	85	134	193
	4	1	1	3	6	9	15	22	35	47	61	98	140
	2	1	1	2	5	6	11	16	25	34	44	70	100
	1	1	1	1	3	5	8	12	19	25	33	53	75
	1/0	0	1	1	3	4	7	10	16	22	28	45	64
	2/0	0	1	1	2	3	6	8	13	18	24	38	54
	3/0	0	1	1	1	3	5	7	11	15	19	31	44
	4/0	0	0	1	1	2	4	6	9	12	16	26	37

250	0	0	1	1	3	5	7	10	13	21	30
300	0	0	1	1	3	4	6	8	11	17	25
350	0	0	1	1	2	3	5	7	10	15	22
400	0	0	0	1	1	3	5	7	9	14	20
500	0	0	0	1	1	2	4	5	7	11	17
600	0	0	1	1	1	1	3	4	6	9	13
700	0	0	0	1	1	1	3	4	5	8	12
750	0	0	0	1	1	1	2	3	5	7	11
1000	0	0	0	0	1	1	1	3	3	6	8

Note: Compact stranding is defined as the result of a manufacturing process where the standard conductor is compressed to the extent that the interstices (voids between strand wires) are virtually eliminated.

Source: NFPA 70, National Electrical Code®, NFPA, Quincy, MA, 2002 edition, Table C9(A).

Exhibit VII.57 Maximum Number of Same Size Conductors or Fixture Wires in Rigid PVC Conduit, Schedule 40 and HDPE Conduit

Conductors

Type	Conductor Size (AWG/kcmil)	Metric Designator (Trade Size)											
		16 (1/2)	21 (3/4)	27 (1)	35 (1-1/4)	41 (1-1/2)	53 (2)	63 (2-1/2)	78 (3)	91 (3-1/2)	103 (4)	129 (5)	155 (6)
RHH, RHW, RHW-2	14	4	7	11	20	27	45	64	99	133	171	269	390
	12	3	5	9	16	22	37	53	82	110	142	224	323
	10	2	4	7	13	18	30	43	66	89	115	181	261
	8	1	2	4	7	9	15	22	35	46	60	94	137
	6	1	1	3	5	7	12	18	28	37	48	76	109
	4	1	1	2	4	6	10	14	22	29	37	59	85
	3	1	1	1	4	5	8	12	19	25	33	52	75
	2	1	1	1	3	4	7	10	16	22	28	45	65
	1	0	1	1	1	3	5	7	11	14	19	29	43
	1/0	0	0	1	1	2	4	6	9	13	16	26	37
	2/0	0	0	1	1	1	3	5	8	11	14	22	32
	3/0	0	0	0	1	1	3	4	7	9	12	19	28
	4/0	0	0	0	1	1	2	4	6	8	10	16	24

250	18	12	8	6	4	3	1	1	1	0	0	0
300	16	11	7	5	4	2	1	1	1	0	0	0
350	14	10	6	5	3	2	1	1	1	0	0	0
400	13	9	6	4	3	1	1	1	1	0	0	0
500	11	8	5	4	3	1	1	0	0	0	0	0
600	9	6	4	3	2	1	1	1	0	0	0	0
700	8	6	3	3	1	1	1	0	0	0	0	0
750	8	5	3	2	1	1	1	0	0	0	0	0
800	7	5	3	2	1	1	1	0	0	0	0	0
900	7	5	3	2	1	1	1	0	0	0	0	0
1000	6	4	3	1	1	1	1	0	0	0	0	0
1250	5	3	1	1	1	1	0	0	0	0	0	0
1500	4	3	1	1	1	1	0	0	0	0	0	0
1750	3	2	1	1	1	1	0	0	0	0	0	0
2000	3	2	1	1	1	0	0	0	0	0	0	0
TW, THHW, THW, THW-2 14	822	568	361	280	209	135	94	57	42	24	14	8
12	631	436	277	215	160	103	72	44	32	18	11	6
10	470	325	206	160	119	77	54	32	24	13	8	4
8	261	181	115	89	66	43	30	18	13	7	4	2

Continued

Exhibit VII.57 Maximum Number of Same Size Conductors or Fixture Wires in Rigid PVC Conduit, Schedule 40 and HDPE Conduit *Continued*

Type	Conductor Size (AWG/kcmil)	Metric Designator (Trade Size)											
		16 (1/2)	21 (3/4)	27 (1)	35 (1-1/4)	41 (1-1/2)	53 (2)	63 (2-1/2)	78 (3)	91 (3-1/2)	103 (4)	129 (5)	155 (6)
RHH,[a] RHW,[a] RHW-2[a]	14	5	9	16	28	38	63	90	139	186	240	378	546
	12	4	8	12	22	30	50	72	112	150	193	304	439
	10	3	6	10	17	24	39	56	87	117	150	237	343
	8	1	3	6	10	14	23	33	52	70	90	142	205
TW, THW, THHW, THW-2	6	1	2	4	8	11	18	26	40	53	69	109	157
	4	1	1	3	6	8	13	19	30	40	51	81	117
	3	1	1	3	5	7	11	16	25	34	44	69	100
	2	1	1	2	4	6	10	14	22	29	37	59	85
	1	0	1	1	3	4	7	10	15	20	26	41	60
	1/0	0	1	1	2	3	6	8	13	17	22	35	51
	2/0	0	1	1	1	3	5	7	11	15	19	30	43
	3/0	0	0	1	1	2	4	6	9	12	16	25	36
	4/0	0	0	0	1	1	3	5	8	10	13	21	30

250	0	0	0	1	1	3	4	6	8	11	17	25
300	0	0	0	1	1	2	3	5	7	9	15	21
350	0	0	1	1	1	2	3	5	6	8	13	19
400	0	0	1	1	1	1	3	4	6	7	12	17
500	0	0	1	1	1	1	2	3	5	6	10	14
600	0	0	0	1	1	1	1	3	4	5	8	11
700	0	0	0	1	1	1	1	2	3	4	7	10
750	0	0	0	1	1	1	1	2	3	4	6	10
800	0	0	0	1	1	1	1	2	3	4	6	9
900	0	0	0	0	1	1	1	1	3	3	6	8
1000	0	0	0	0	1	1	1	1	2	3	5	7
1250	0	0	0	0	1	1	1	1	1	2	4	6
1500	0	0	0	0	0	1	1	1	1	1	3	5
1750	0	0	0	0	0	0	1	1	1	1	3	4
2000	0	0	0	0	0	0	1	1	1	1	3	4
14	11	21	34	60	82	135	193	299	401	517	815	1178
12	8	15	25	43	59	99	141	218	293	377	594	859
10	5	9	15	27	37	62	89	137	184	238	374	541
8	3	5	9	16	21	36	51	79	106	137	216	312
6	1	4	6	11	15	26	37	57	77	99	156	225

THHN, THWN, THWN-2

Continued

Exhibit VII.57 Maximum Number of Same Size Conductors or Fixture Wires in Rigid PVC Conduit, Schedule 40 and HDPE Conduit *Continued*

				Metric Designator (Trade Size)									
Type	Conductor Size (AWG/kcmil)	16 (1/2)	21 (3/4)	27 (1)	35 (1-1/4)	41 (1-1/2)	53 (2)	63 (2-1/2)	78 (3)	91 (3-1/2)	103 (4)	129 (5)	155 (6)
THHN, THWN, THWN-2	4	1	2	4	7	9	16	22	35	47	61	96	138
	3	1	1	3	6	8	13	19	30	40	51	81	117
	2	1	1	3	5	7	11	16	25	33	43	68	98
	1	1	1	1	3	5	8	12	18	25	32	50	73
	1/0	0	1	1	3	4	7	10	15	21	27	42	61
	2/0	0	1	1	2	3	6	8	13	17	22	35	51
	3/0	0	1	1	1	3	5	7	11	14	18	29	42
	4/0	0	0	1	1	2	4	6	9	12	15	24	35
	250	0	0	1	1	1	3	4	7	10	12	20	28
	300	0	0	1	1	1	3	4	6	8	11	17	24
	350	0	0	0	1	1	2	3	5	7	9	15	21
	400	0	0	0	1	1	1	3	5	6	8	13	19
	500	0	0	0	1	1	1	2	4	5	7	11	16

Type	Size												
	600	0	0	0	1	1	1	1	3	4	5	9	13
	700	0	0	0	0	1	1	1	3	4	5	8	11
	750	0	0	0	0	1	1	1	2	3	4	7	11
	800	0	0	0	0	1	1	1	2	3	4	7	10
	900	0	0	0	0	1	1	1	2	3	4	6	9
	1000	0	0	0	0	0	1	1	1	3	3	6	8
FEP, FEPB, PFA, PFAH, TFE	14	11	20	33	58	79	131	188	290	389	502	790	1142
	12	8	15	24	42	58	96	137	212	284	366	577	834
	10	6	10	17	30	41	69	98	152	204	263	414	598
	8	3	6	10	17	24	39	56	87	117	150	237	343
	6	2	4	7	12	17	28	40	62	83	107	169	244
	4	1	3	5	8	12	19	28	43	58	75	118	170
	3	1	2	4	7	10	16	23	36	48	62	98	142
	2	1	1	3	6	8	13	19	30	40	51	81	117
PFA, PFAH, TFE	1	1	1	2	4	5	9	13	20	28	36	56	81
PFA, PFAH, TFE, Z	1/0	1	1	1	3	4	8	11	17	23	30	47	68
	2/0	0	1	1	3	4	6	9	14	19	24	39	56
	3/0	0	1	1	2	3	5	7	12	16	20	32	46
	4/0	0	1	1	1	2	4	6	9	13	16	26	38

Continued

Exhibit VII.57 Maximum Number of Same Size Conductors or Fixture Wires in Rigid PVC Conduit, Schedule 40 and HDPE Conduit *Continued*

Conductors

Type	Conductor Size (AWG/kcmil)	16 (1/2)	21 (3/4)	27 (1)	35 (1-1/4)	41 (1-1/2)	53 (2)	63 (2-1/2)	78 (3)	91 (3-1/2)	103 (4)	129 (5)	155 (6)
							Metric Designator (Trade Size)						
Z	14	13	24	40	70	95	158	226	350	469	605	952	1376
	12	9	17	28	49	68	112	160	248	333	429	675	976
	10	6	10	17	30	41	69	98	152	204	263	414	598
	8	3	6	11	19	26	43	62	96	129	166	261	378
	6	2	4	7	13	18	30	43	67	90	116	184	265
	4	1	3	5	9	12	21	30	46	62	80	126	183
	3	1	2	4	6	9	15	22	34	45	58	92	133
	2	1	1	3	5	7	12	18	28	38	49	77	111
	1	1	1	2	4	6	10	14	23	30	39	62	90
XHH, XHHW, XHHW-2, ZW	14	8	14	24	42	57	94	135	209	280	361	568	822
	12	6	11	18	32	44	72	103	160	215	277	436	631
	10	4	8	13	24	32	54	77	119	160	206	325	470
	8	2	4	7	13	18	30	43	66	89	115	181	261

Type	Size												
XHH, XHHW, XHHW-2	6	193	134	85	66	49	32	22	13	10	5	3	1
	4	140	97	61	48	35	23	16	9	7	4	2	1
	3	118	82	52	40	30	19	13	8	6	3	1	1
	2	99	69	44	34	25	16	11	7	5	3	1	1
	1	74	51	32	25	19	12	8	5	3	1	1	1
	1/0	62	43	27	21	16	10	7	4	3	1	1	0
	2/0	52	36	23	17	13	8	6	3	2	1	1	0
	3/0	43	30	19	14	11	7	5	3	1	1	1	1
	4/0	35	24	15	12	9	6	4	2	1	1	1	0
	250	29	20	13	10	7	5	3	1	1	1	0	0
	300	25	17	11	8	6	4	3	1	1	1	0	0
	350	22	15	9	7	5	3	2	1	1	0	0	0
	400	19	13	8	6	5	3	1	1	1	0	0	0
	500	16	11	7	5	4	2	1	1	1	0	0	0
	600	13	9	5	4	3	1	1	1	0	0	0	0
	700	11	8	5	4	3	1	1	1	0	0	0	0
	750	11	7	4	3	2	1	1	1	0	0	0	0
	800	10	7	4	3	2	1	1	1	0	0	0	0
	900	9	6	4	3	2	1	1	1	0	0	0	0

Continued

Exhibit VII.57 Maximum Number of Same Size Conductors or Fixture Wires in Rigid PVC Conduit, Schedule 40 and HDPE Conduit *Continued*

Type	Conductor Size (AWG/kcmil)	Conductors — Metric Designator (Trade Size)											
		16 (1/2)	21 (3/4)	27 (1)	35 (1-1/4)	41 (1-1/2)	53 (2)	63 (2-1/2)	78 (3)	91 (3-1/2)	103 (4)	129 (5)	155 (6)
XHH, XHHW, XHHW-2	1000	0	0	0	0	0	1	1	1	3	3	6	8
	1250	0	0	0	0	0	1	1	1	1	3	4	6
	1500	0	0	0	0	0	1	1	1	1	2	4	5
	1750	0	0	0	0	0	0	1	1	1	1	3	5
	2000	0	0	0	0	0	0	1	1	1	1	3	4

Fixture Wires

Type	Conductor Size (AWG/kcmil)	Metric Designator (Trade Size)						
		16 (1/2)	21 (3/4)	27 (1)	35 (1-1/4)	41 (1-1/2)	53 (2)
FFH-2, RFH-2, RFHH-3	18	8	14	23	40	54	90
	16	6	12	19	33	46	76
SF-2, SFF-2	18	10	17	29	50	69	114
	16	8	14	24	42	57	94
	14	6	12	19	33	46	76
SF-1, SFF-1	18	17	31	51	89	122	202
RFH-1, RFHH-2, TF, TFF, XF, XFF	18	13	23	38	66	90	149
RFHH-2, TF, TFF, XF, XFF	16	10	18	30	53	73	120

Continued

Exhibit VII.57 Maximum Number of Same Size Conductors or Fixture Wires in Rigid PVC Conduit, Schedule 40 and HDPE Conduit *Continued*

Fixture Wires

Type	Conductor Size (AWG/kcmil)	Metric Designator (Trade Size)							
		16 (1/2)	21 (3/4)	27 (1)	35 (1-1/4)	41 (1-1/2)	53 (2)
XF, XFF	14	8	14	24	42	57	94
TFN, TFFN	18	20	37	60	105	144	239
	16	16	28	46	80	110	183
PF, PFF, PGF, PGFF, PAF, PTF, PTFF, PAFF	18	19	35	57	100	137	227
	16	15	27	44	77	106	175
	14	11	20	33	58	79	131
HF, HFF, ZF, ZFF, ZHF	18	25	45	74	129	176	292
	16	18	33	54	95	130	216
	14	13	24	40	70	95	158
KF-2, KFF-2	18	36	65	107	187	256	424
	16	26	46	75	132	180	299
	14	17	31	52	90	124	205

| | 12 | 12 | 22 | 35 | 62 | 85 | 141 |
	10	8	14	24	42	57	94
KF-1, KFF-1,	18	43	78	128	223	305	506
	16	30	55	90	157	214	355
	14	20	37	60	105	144	239
	12	13	24	40	70	95	158
	10	9	16	26	45	62	103
XF, XFF	12	4	8	12	22	30	50
	10	3	6	10	17	24	39

Note: This table is for concentric stranded Conductors only. For compact stranded Conductors, Exhibit VII.58 should be used.

[a]Types RHH, RHW, and RHW-2 without outer covering.

Source: NFPA 70, *National Electrical Code®*, NFPA, Quincy, MA, 2002 edition, Table C10.

Exhibit VII.58 Maximum Number of Same Size Compact Conductors in Rigid PVC Conduit, Schedule 40 and HDPE Conduit

Compact Conductors

Type	Conductor Size (AWG/kcmil)	Metric Designator (Trade Size)											
		16 (1/2)	21 (3/4)	27 (1)	35 (1-1/4)	41 (1-1/2)	53 (2)	63 (2-1/2)	78 (3)	91 (3-1/2)	103 (4)	129 (5)	155 (6)
THW, THW-2, THHW	8	1	3	6	11	15	26	37	57	76	98	155	224
	6	1	3	5	9	12	20	28	44	59	76	119	173
	4	1	1	3	6	9	15	21	33	44	57	89	129
	2	1	1	2	5	6	11	15	24	32	42	66	95
	1	1	1	1	3	4	7	11	17	23	29	46	67
	1/0	0	1	1	3	4	6	9	15	20	25	40	58
	2/0	0	1	1	2	3	5	8	12	16	21	34	49
	3/0	0	1	1	1	3	5	7	10	14	18	29	42
	4/0	0	0	1	1	2	4	5	9	12	15	24	35
	250	0	0	1	1	1	3	4	7	9	12	19	27
	300	0	0	1	1	1	2	4	6	8	10	16	24
	350	0	0	1	1	1	2	3	5	7	9	15	21
	400	0	0	0	1	1	1	3	5	6	8	13	19
	500	0	0	0	1	1	1	2	4	5	7	11	16

THHN, THWN, THWN-2												
600	13	9	5	4	3	1	1	1	1	0	0	0
700	12	8	5	4	3	1	1	1	0	0	0	0
750	11	7	5	3	2	1	1	1	0	0	0	0
1000	9	6	4	3	1	1	1	1	0	0	0	0
8	—	—	—	—	—	—	—	—	—	—	—	—
6	253	175	111	86	64	41	29	17	13	7	4	2
4	156	108	68	53	40	25	18	11	8	4	2	1
2	112	77	49	38	28	18	13	8	5	3	1	1
1	84	58	37	29	21	14	9	6	4	2	1	1
1/0	72	49	31	24	18	12	8	5	3	1	1	1
2/0	59	41	26	20	15	9	7	4	3	1	1	0
3/0	50	34	22	17	12	8	5	3	2	1	1	0
4/0	41	28	18	14	10	6	4	3	1	1	1	0
250	32	22	14	11	8	5	3	1	1	1	0	0
300	28	19	12	9	7	4	3	1	1	1	0	0
350	24	17	10	8	6	4	3	1	1	1	0	0
400	22	15	9	7	5	3	2	1	1	1	0	0
500	18	13	8	6	4	3	1	1	1	0	0	0

Continued

Exhibit VII.58 Maximum Number of Same Size Compact Conductors in Rigid PVC Conduit, Schedule 40 and HDPE Conduit Continued

Compact Conductors

Type	Conductor Size (AWG/kcmil)	16 (1/2)	21 (3/4)	27 (1)	35 (1-1/4)	41 (1-1/2)	53 (2)	63 (2-1/2)	78 (3)	91 (3-1/2)	103 (4)	129 (5)	155 (6)
THHN, THWN, THWN-2	600	0	0	0	1	1	1	2	4	5	6	10	15
	700	0	0	0	1	1	1	1	3	4	5	9	13
	750	0	0	0	1	1	1	1	3	4	5	8	12
	1000	0	0	0	0	1	1	1	2	3	4	6	9
XHHW, XHHW-2	8	3	5	8	14	20	33	47	73	99	127	200	290
	6	1	3	6	11	15	25	35	55	73	94	149	215
	4	1	2	4	8	11	18	25	40	53	68	108	156
	2	1	1	3	5	8	13	18	28	38	49	77	112
	1	1	1	2	4	6	10	14	21	29	37	58	84
	1/0	1	1	1	3	5	8	12	18	24	31	49	72
	2/0	1	1	1	3	4	7	10	15	20	26	42	60
	3/0	0	1	1	2	3	5	8	12	17	22	34	50
	4/0	0	1	1	1	3	5	7	10	14	18	29	42

250	0	0	1	1	4	5	8	11	14	23	33
300	0	0	1	1	3	4	7	9	12	19	28
350	0	0	1	1	3	4	6	8	11	17	25
400	0	0	1	1	2	3	5	7	10	15	22
500	0	0	0	1	1	3	4	6	8	13	18
600	0	0	0	1	1	2	4	5	6	10	15
700	0	0	0	1	1	1	3	4	5	9	13
750	0	0	0	1	1	1	3	4	5	8	12
1000	0	0	1	1	1	1	2	3	4	6	9

Note: Compact stranding is defined as the result of a manufacturing process where the standard conductor is compressed to the extent that the interstices (voids between strand wires) are virtually eliminated:

Source: NFPA 70, National Electrical Code®, NFPA, Quincy, MA, 2002 edition, Table C10(A).

Exhibit VII.59 Maximum Number of Same Size Conductors or Fixture Wires in Type A, Rigid PVC Conduit

					Conductors					
					Metric Designator (Trade Size)					
Type	Conductor Size (AWG/kcmil)	16 (1/2)	21 (3/4)	27 (1)	35 (1-1/4)	41 (1-1/2)	53 (2)	63 (2-1/2)	78 (3)	91 (3-1/2)	103 (4)
RHH, RHW, RHW-2	14	5	9	15	24	31	49	74	112	146	187
	12	4	7	12	20	26	41	61	93	121	155
	10	3	6	10	16	21	33	50	75	98	125
	8	1	3	5	8	11	17	26	39	51	65
	6	1	2	4	6	9	14	21	31	41	52
	4	1	1	3	5	7	11	16	24	32	41
	3	1	1	3	4	6	9	14	21	28	36
	2	1	1	2	4	5	8	12	18	24	31
	1	0	1	1	2	3	5	8	12	16	20
	1/0	0	1	1	2	3	5	7	10	14	18
	2/0	0	1	1	1	2	4	6	9	12	15
	3/0	0	0	1	1	1	3	5	8	10	13
	4/0	0	0	1	1	1	3	4	7	9	11

Size										
250	0	0	1	1	1	1	3	5	7	8
300	0	0	1	1	1	1	3	4	6	7
350	0	0	0	1	1	1	2	4	5	7
400	0	0	0	1	1	1	2	4	5	6
500	0	0	0	1	1	1	1	3	4	5
600	0	0	0	0	1	1	1	2	3	4
700	0	0	0	0	1	1	1	2	3	4
750	0	0	0	0	1	1	1	1	3	4
800	0	0	0	0	0	1	1	1	3	3
900	0	0	0	0	0	0	1	1	2	3
1000	0	0	0	0	0	1	1	1	2	3
1250	0	0	0	0	0	0	1	1	1	2
1500	0	0	0	0	0	0	1	1	1	1
1750	0	0	0	0	0	0	1	1	1	1
2000	0	0	0	0	0	0	1	1	1	1
TW, THHW, THW, THW-2 14	11	18	31	51	67	105	157	235	307	395
12	8	14	24	39	51	80	120	181	236	303
10	6	10	18	29	38	60	89	135	176	226
8	3	6	10	16	21	33	50	75	98	125

Continued

Exhibit VII.59 Maximum Number of Same Size Conductors or Fixture Wires in Type A, Rigid PVC Conduit *Continued*

Conductors

Type	Conductor Size (AWG/kcmil)	Metric Designator (Trade Size)									
		16 (1/2)	21 (3/4)	27 (1)	35 (1-1/4)	41 (1-1/2)	53 (2)	63 (2-1/2)	78 (3)	91 (3-1/2)	103 (4)
RHH,[a] RHW,[a] RHW-2[a]	14	7	12	20	34	44	70	104	157	204	262
	12	6	10	16	27	35	56	84	126	164	211
	10	4	8	13	21	28	44	65	98	128	165
	8	2	4	8	12	16	26	39	59	77	98
RHH, RHW[a]	6	1	3	6	9	13	20	30	45	59	75
TW, THW, THHW, THW-2	4	1	2	4	7	9	15	22	33	44	56
	3	1	1	4	6	8	13	19	29	37	48
	2	1	1	3	5	7	11	16	24	32	41
	1	1	1	1	3	5	7	11	17	22	29
	1/0	1	1	1	3	4	6	10	14	19	24
	2/0	0	1	1	2	3	5	8	12	16	21
	3/0	0	1	1	1	3	4	7	10	13	17
	4/0	0	1	1	1	2	4	6	9	11	14

250	12	9	7	4	3	1	1	1	0	0
300	10	8	6	4	2	1	1	1	0	0
350	9	7	5	3	2	1	1	1	0	0
400	8	6	5	3	1	1	1	1	0	0
500	7	5	4	2	1	1	1	0	0	0
600	5	4	3	1	1	1	1	1	0	0
700	5	4	3	1	1	1	1	1	0	0
750	4	3	3	1	1	1	1	0	0	0
800	4	3	2	1	1	1	0	0	0	0
900	4	3	2	1	1	1	0	0	0	0
1000	3	3	1	1	1	1	0	0	0	0
1250	3	1	1	1	1	0	0	0	0	0
1500	2	1	1	1	1	0	0	0	0	0
1750	1	1	1	1	0	0	0	0	0	0
2000	1	1	1	1	0	0	0	0	0	0
THHN, THWN, THWN-2 14	566	441	338	225	150	96	73	44	27	16
12	412	321	246	164	109	70	53	32	19	11
10	260	202	155	103	69	44	33	20	12	7
8	150	117	89	59	40	25	19	12	7	4
6	108	84	64	43	28	18	14	8	5	3

Continued

Exhibit VII.59 Maximum Number of Same Size Conductors or Fixture Wires in Type A, Rigid PVC Conduit *Continued*

		Conductors									
	Conductor Size (AWG/kcmil)	Metric Designator (Trade Size)									
Type		16 (1/2)	21 (3/4)	27 (1)	35 (1-1/4)	41 (1-1/2)	53 (2)	63 (2-1/2)	78 (3)	91 (3-1/2)	103 (4)
THHN, THWN, THWN-2	4	1	3	5	8	11	17	26	39	52	66
	3	1	2	4	7	9	15	22	33	44	56
	2	1	1	3	6	8	12	19	28	37	47
	1	1	1	2	4	6	9	14	21	27	35
	1/0	1	1	2	4	5	8	11	17	23	29
	2/0	1	1	1	3	4	6	10	14	19	24
	3/0	0	1	1	2	3	5	8	12	16	20
	4/0	0	1	1	1	3	4	6	10	13	17
	250	0	1	1	1	2	3	5	8	10	14
	300	0	0	1	1	1	3	4	7	9	12
	350	0	0	1	1	1	2	4	6	8	10
	400	0	0	0	1	1	2	3	5	7	9
	500	0	0	0	1	1	1	3	4	6	7
	600	0	0	0	1	1	1	2	3	5	6
	700	0	0	0	0	1	1	1	3	4	5

Continued

Type	Size										
	750	0	0	0	0	1	1	1	3	4	5
	800	0	0	0	1	1	1	1	3	4	5
	900	0	0	0	1	1	1	1	2	3	4
	1000	0	0	0	0	1	1	1	2	3	4
FEP, FEPB, PFA, PFAH, TFE	14	15	26	43	70	93	146	218	327	427	549
	12	11	19	31	51	68	106	159	239	312	400
	10	8	13	22	37	48	76	114	171	224	287
	8	4	8	13	21	28	44	65	98	128	165
	6	3	5	9	15	20	31	46	70	91	117
	4	1	4	6	10	14	21	32	49	64	82
	3	1	3	5	8	11	18	27	40	53	68
	2	1	2	4	7	9	15	22	33	44	56
PFA, PFAH, TFE	1	1	1	3	5	6	10	15	23	30	39
PFA, PFAH, TFE, Z	1/0	1	1	2	4	5	8	13	19	25	32
	2/0	1	1	1	3	4	7	10	16	21	27
	3/0	1	1	1	3	3	6	9	13	17	22
	4/0	0	1	1	2	3	5	7	11	14	18
Z	14	18	31	52	85	112	175	263	395	515	661
	12	13	22	37	60	79	124	186	280	365	469

Exhibit VII.59 Maximum Number of Same Size Conductors or Fixture Wires in Type A, Rigid PVC Conduit *Continued*

Conductors

Type	Conductor Size (AWG/kcmil)	Metric Designator (Trade Size)									
		16 (1/2)	21 (3/4)	27 (1)	35 (1-1/4)	41 (1-1/2)	53 (2)	63 (2-1/2)	78 (3)	91 (3-1/2)	103 (4)
Z	10	8	13	22	37	48	76	114	171	224	287
	8	5	8	14	23	30	48	72	108	141	181
	6	3	6	10	16	21	34	50	76	99	127
	4	2	4	7	11	15	23	35	52	68	88
	3	1	3	5	8	11	17	25	38	50	64
	2	1	2	4	7	9	14	21	32	41	53
	1	1	1	3	5	7	11	17	26	33	43
XHH, XHHW, XHHW-2, ZW	14	11	18	31	51	67	105	157	235	307	395
	12	8	14	24	39	51	80	120	181	236	303
	10	6	10	18	29	38	60	89	135	176	226
	8	3	6	10	16	21	33	50	75	98	125
	6	2	4	7	12	15	24	37	55	72	93
	4	1	3	5	8	11	18	26	40	52	67
	3	1	2	4	7	9	15	22	34	44	57
	2	1	1	3	6	8	12	19	28	37	48

XHH, XHHW, XHHW-2

Size	35	28	21	14	9	6	4	3	1	1
1	35	28	21	14	9	6	4	3	1	1
1/0	30	23	18	12	8	5	4	2	1	1
2/0	25	19	15	10	6	4	3	2	1	1
3/0	20	16	12	8	5	3	2	1	1	0
4/0	17	13	10	7	4	3	1	1	1	0
250	14	11	8	5	3	2	1	1	0	0
300	12	9	7	5	3	1	1	1	0	0
350	10	8	6	4	3	1	1	1	0	0
400	9	7	5	3	2	1	1	1	0	0
500	8	6	4	3	1	1	1	1	0	0
600	6	5	3	2	1	1	1	0	0	0
700	5	4	3	1	1	1	1	0	0	0
750	5	4	3	1	1	1	1	0	0	0
800	5	4	3	1	1	1	1	0	0	0
900	4	3	2	1	1	1	0	0	0	0
1000	4	3	2	1	1	1	0	0	0	0
1250	3	2	1	1	1	0	0	0	0	0
1500	2	1	1	1	1	0	0	0	0	0
1750	2	1	1	1	1	0	0	0	0	0
2000	1	1	1	1	0	0	0	0	0	0

Continued

Exhibit VII.59 Maximum Number of Same Size Conductors or Fixture Wires in Type A, Rigid PVC Conduit *Continued*

Fixture Wires

Type	Conductor Size (AWG/kcmil)	16 (1/2)	21 (3/4)	27 (1)	35 (1-1/4)	41 (1-1/2)	53 (2)
				Metric Designator (Trade Size)			
FFH-2, RFH-2, RFHH-3	18	10	18	30	48	64	100
	16	9	15	25	41	54	85
SF-2, SFF-2	18	13	22	37	61	81	127
	16	11	18	31	51	67	105
	14	9	15	25	41	54	85
SF-1, SFF-1	18	23	40	66	108	143	224
RFH-1, RFHH-2, TF, TFF, XF, XFF	18	17	29	49	80	105	165
RFHH-2, TF, TFF, XF, XFF	16	14	24	39	65	85	134
XF, XFF	14	11	18	31	51	67	105
TFN, TFFN	18	28	47	79	128	169	265
	16	21	36	60	98	129	202
PF, PFF, PGF	18	26	45	74	122	160	251

PGFF, PAF, PTF, PTFF, PAFF	16	20	34	58	94	124	194
	14	15	26	43	70	93	146
HF, HFF, ZF, ZFF, ZHF	18	34	58	96	157	206	324
	16	25	42	71	116	152	239
	14	18	31	52	85	112	175
KF-2, KFF-2	18	49	84	140	228	300	470
	16	35	59	98	160	211	331
	14	24	40	67	110	145	228
	12	16	28	46	76	100	157
	10	11	18	31	51	67	105
KF-1, KFF-1	18	59	100	167	272	357	561
	16	41	70	117	191	251	394
	14	28	47	79	128	169	265
	12	18	31	52	85	112	175
	10	12	20	34	55	73	115
XF, XFF	12	6	10	16	27	35	56
	10	4	8	13	21	28	44

Note: This table is for concentric stranded Conductors only. For compact stranded Conductors, Exhibit VII.60 should be used.
[a]Types RHH, RHW, and RWH-2 without outer covering.
Source: NFPA 70, *National Electrical Code®*, NFPA, Quincy, MA, 2002 edition, Table C11.

Exhibit VII.60 Maximum Number of Same Size Compact Conductors in Type A, Rigid PVC Conduit

Compact Conductors

Type	Conductor Size (AWG/kcmil)	16 (1/2)	21 (3/4)	27 (1)	35 (1-1/4)	41 (1-1/2)	53 (2)	63 (2-1/2)	78 (3)	91 (3-1/2)	103 (4)
THW, THW-2, THHW	8	3	5	8	14	18	28	42	64	84	107
	6	2	4	6	10	14	22	33	49	65	83
	4	1	3	5	8	10	16	24	37	48	62
	2	1	1	3	6	7	12	18	27	36	46
	1	1	1	2	4	5	8	13	19	25	32
	1/0	1	1	1	3	4	7	11	16	21	28
	2/0	1	1	1	3	4	6	9	14	18	23
	3/0	0	1	1	2	3	5	8	12	15	20
	4/0	0	1	1	1	3	4	6	10	13	17
	250	0	1	1	1	1	3	5	8	10	13
	300	0	0	1	1	1	3	4	7	9	11
	350	0	0	1	1	1	3	4	6	8	10
	400	0	0	1	1	1	2	3	5	7	9
	500	0	0	1	1	1	1	3	4	6	8

Size										
600	6	5	3	2	1	1	1	0	0	0
700	5	4	3	1	1	1	1	0	0	0
750	5	4	3	1	1	1	1	0	0	0
1000	4	3	2	1	1	1	0	0	0	0
THHN, THWN, THWN-2										
8	—	—	—	—	—	—	—	—	—	—
6	121	94	72	48	32	20	15	9	5	3
4	75	58	45	30	20	12	9	6	3	1
2	54	42	32	21	14	9	7	4	2	1
1	40	31	24	16	10	7	5	3	1	1
1/0	34	27	20	13	9	6	4	2	1	1
2/0	28	22	17	11	7	5	3	1	1	1
3/0	24	18	14	9	6	4	3	1	1	1
4/0	19	15	11	8	5	3	2	1	1	0
250	15	12	9	6	4	2	1	1	1	0
300	13	10	8	5	3	1	1	1	1	0
350	11	9	7	4	3	1	1	1	0	0
400	10	8	6	4	2	1	1	1	1	0
500	9	7	5	3	2	1	1	1	1	0
600	7	5	4	3	1	1	1	0	0	0
700	6	5	3	2	1	1	1	0	0	0

Continued

Exhibit VII.60 Maximum Number of Same Size Compact Conductors in Type A, Rigid PVC Conduit *Continued*

Compact Conductors

Type	Conductor Size (AWG/kcmil)	Metric Designator (Trade Size)									
		16 (1/2)	21 (3/4)	27 (1)	35 (1-1/4)	41 (1-1/2)	53 (2)	63 (2-1/2)	78 (3)	91 (3-1/2)	103 (4)
THHN, THWN, THWN-2	750	0	0	0	1	1	1	2	3	4	6
	1000	0	0	0	0	1	1	2	2	3	4
XHHW, XHHW-2	8	4	6	11	18	23	37	55	83	108	139
	6	3	5	8	13	17	27	41	62	80	103
	4	1	3	6	9	12	20	30	45	58	75
	2	1	2	4	7	9	14	21	32	42	54
	1	1	1	3	5	7	10	16	24	31	40
	1/0	1	1	2	4	6	9	13	20	27	34
	2/0	1	1	1	3	5	7	11	17	22	29
	3/0	1	1	1	3	4	6	9	14	18	24
	4/0	0	1	1	2	3	5	8	12	15	20
	250	0	1	1	1	2	4	6	9	12	16
	300	0	1	1	1	1	3	5	8	10	13
	350	0	0	1	1	1	3	5	7	9	12

400	0	0	0	1	1	1	3	4	6	8	11
500	0	0	0	1	1	1	2	3	5	7	9
600	0	0	0	0	1	1	1	3	4	5	7
700	0	0	0	0	1	1	1	2	3	5	6
750	0	0	0	0	1	1	1	2	3	4	6
1000	0	0	0	0	0	1	1	1	2	3	4

Note: Compact stranding is defined as the result of a manufacturing process where the standard conductor is compressed to the extent that the interstices (voids between strand wires) are virtually eliminated.

Source: NFPA 70, *National Electrical Code®*, NFPA, Quincy, MA, 2002 edition, Table C11(A).

Exhibit VII.61 Maximum Number of Same Size Conductors in Type EB, PVC Conduit

Type	Conductor Size (AWG/kcmil)	Conductors						
		Metric Designator (Trade Size)						
		53 (2)	78 (3)	91 (3-1/2)	103 (4)	129 (5)	155 (6)	
RHH, RHW,	14	53	119	155	197	303	430	
RHW-2	12	44	98	128	163	251	357	
RH, RHH, RHW,	10	35	79	104	132	203	288	
RHW-2	8	18	41	54	69	106	151	
	6	15	33	43	55	85	121	
	4	11	26	34	43	66	94	
	3	10	23	30	38	58	83	
	2	9	20	26	33	50	72	
	1	6	13	17	21	33	47	
	1/0	5	11	15	19	29	41	
	2/0	4	10	13	16	25	36	
	3/0	4	8	11	14	22	31	
	4/0	3	7	9	12	18	26	
	250	2	5	7	9	14	20	
	300	1	5	6	8	12	17	

Type	Size						
	350	1		4	5	7	11	16
	400	1		4	5	6	10	14
	500	1		3	4	5	9	12
	600	1	1	3	3	4	7	10
	700	1	1	2	3	4	6	9
	750	1	1	2	3	4	6	9
	800	1	1	2	3	4	6	8
	900	1	1	1	2	3	5	7
	1000	1	1	1	2	3	5	7
	1250	1	1	1	1	2	3	5
	1500	0	1	1	1	1	3	4
	1750	0	1	1	1	1	3	4
	2000	0	1	1	1	1	2	3
TW, THHW, THW, THW-2	14	111	250	327	415	638	907
	12	85	192	251	319	490	696
	10	63	143	187	238	365	519
	8	35	79	104	132	203	288
RHH,[a] RHW,[a] WH-2[a]	14	74	166	217	276	424	603
RHH,[a] HW,[a] RHW-2[a]	12	59	134	175	222	341	485
	10	46	104	136	173	266	378

Continued

Exhibit VII.61 Maximum Number of Same Size Conductors in Type EB, PVC Conduit *Continued*

	Conductors						
	Conductor Size (AWG/kcmil)	Metric Designator (Trade Size)					
Type		53 (2)	78 (3)	91 (3-1/2)	103 (4)	129 (5)	155 (6)
RHH,[a] HW,[a] RHW-2[a]	8	28	62	81	104	159	227
RHH,[a] RHW,[a] RHW-2,[a] TW, THW, THHW, THW-2	6	21	48	62	79	122	173
	4	16	36	46	59	91	129
	3	13	30	40	51	78	111
	2	11	26	34	43	66	94
	1	8	18	24	30	46	66
	1/0	7	15	20	26	40	56
	2/0	6	13	17	22	34	48
	3/0	5	11	14	18	28	40
	4/0	4	9	12	15	24	34
	250	3	7	10	12	19	27
	300	3	6	8	11	17	24
	350	2	6	7	9	15	21
	400	2	5	7	8	13	19
	500	1	4	5	7	11	16

600	1		3	4	6	9	13
700	1		3	4	5	8	11
750	1		3	4	5	7	11
800	1		3	3	4	7	10
900	1		2	3	4	6	9
1000	1		2	3	4	6	8
1250	1	1	1	2	3	4	6
1500	1	1	1	1	2	4	6
1750	1	1	1	1	2	3	5
2000	0	1	1	1	1	3	4

THHN, THWN, THWN-2

14	159	359	468	595	915	1300
12	116	262	342	434	667	948
10	73	165	215	274	420	597
8	42	95	124	158	242	344
6	30	68	89	114	175	248
4	19	42	55	70	107	153
3	16	36	46	59	91	129
2	13	30	39	50	76	109
1	10	22	29	37	57	80

Continued

CONDUIT AND RACEWAY SYSTEMS

Exhibit VII.61 Maximum Number of Same Size Conductors in Type EB, PVC Conduit *Continued*

Type	Conductor Size (AWG/kcmil)	Conductors — Metric Designator (Trade Size)					
		53 (2)	78 (3)	91 (3-1/2)	103 (4)	129 (5)	155 (6)
THHN, THWN, THWN-2	1/0	8	18	24	31	48	68
	2/0	7	15	20	26	40	56
	3/0	5	13	17	21	33	47
	4/0	4	10	14	18	27	39
	250	4	8	11	14	22	31
	300	3	7	10	12	19	27
	350	3	6	8	11	17	24
	400	2	6	7	10	15	21
	500	1	5	6	8	12	18
	600	1	4	5	6	10	14
	700	1	3	4	6	9	12
	750	1	3	4	5	8	12
	800	1	3	4	5	8	11
	900	1	3	3	4	7	10
	1000	1	2	3	4	6	9

FEP, FEPB, PFA, PFAH, TFE	14	155	348	454	578	888	1261
	12	113	254	332	422	648	920
	10	81	182	238	302	465	660
	8	46	104	136	173	266	378
	6	33	74	97	123	189	269
	4	23	52	68	86	132	188
	3	19	43	56	72	110	157
	2	16	36	46	59	91	129
PFA, PFAH, TFE	1	11	25	32	41	63	90
PFA, PFAH, TFE, Z	1/0	9	20	27	34	53	75
	2/0	7	17	22	28	43	62
	3/0	6	14	18	23	36	51
	4/0	5	11	15	19	29	42
Z	14	186	419	547	696	1069	1519
	12	132	297	388	494	759	1078
	10	81	182	238	302	465	660
	8	51	115	150	191	294	417
	6	36	81	105	134	206	293
	4	24	55	72	92	142	201
	3	18	40	53	67	104	147

Continued

Exhibit VII.61 Maximum Number of Same Size Conductors in Type EB, PVC Conduit *Continued*

Conductors

Type	Conductor Size (AWG/kcmil)	Metric Designator (Trade Size)					
		53 (2)	78 (3)	91 (3-1/2)	103 (4)	129 (5)	155 (6)
Z	2	15	34	44	56	86	122
	1	12	27	36	45	70	99
XHH, XHHW, XHHW-2, ZW	14	111	250	327	415	638	907
	12	85	192	251	319	490	696
	10	63	143	187	238	365	519
	8	35	79	104	132	203	288
	6	26	59	77	98	150	213
	4	19	42	56	71	109	155
	3	16	36	47	60	92	131
	2	13	30	39	50	77	110
XHH, XHHW, XHHW-2	1	10	22	29	37	58	82
	1/0	8	19	25	31	48	69
	2/0	7	16	20	26	40	57
	3/0	6	13	17	22	33	47
	4/0	5	11	14	18	27	39

Size						
250	4	9	11	15	22	32
300	3	7	10	12	19	28
350	3	6	9	11	17	24
400	2	6	8	10	15	22
500	1	5	6	8	12	18
600	1	4	5	6	10	14
700	1	3	4	6	9	12
750	1	3	4	5	8	12
800	1	3	4	5	8	11
900	1	3	3	4	7	10
1000	1	2	3	4	6	9
1250	1	1	2	3	5	7
1500	1	1	1	3	4	6
1750	1	1	1	2	4	5
2000	0	1	1	1	3	5

Note: This table is for concentric stranded Conductors only. For compact stranded Conductors, Exhibit VII.62 should be used.

[a]Types RHH, RHW, and RHW-2 without outer covering.

Source: NFPA 70, *National Electrical Code®*, NFPA, Quincy, MA, 2002 edition, Table C12.

Exhibit VII.62 Maximum Number of Same Size Compact Conductors in Type EB, PVC Conduit

Compact Conductors

Type	Conductor Size (AWG/kcmil)	Metric Designator (Trade Size)					
		53 (2)	78 (3)	91 (3-1/2)	103 (4)	129 (5)	155 (6)
THW, THW-2, THHW	8	30	68	89	113	174	247
	6	23	52	69	87	134	191
	4	17	39	51	65	100	143
	2	13	29	38	48	74	105
	1	9	20	26	34	52	74
	1/0	8	17	23	29	45	64
	2/0	6	15	19	24	38	54
	3/0	5	12	16	21	32	46
	4/0	4	10	14	17	27	38
	250	3	8	11	14	21	30
	300	3	7	9	12	19	26
	350	3	6	8	11	17	24
	400	2	6	7	10	15	21
	500	1	5	6	8	12	18

600	1	4	5	6	10	14
700	1	3	4	6	9	13
750	1	3	4	5	8	12
1000	1	2	3	4	7	9
8	—	—	—	—	—	—
6	34	77	100	128	196	279
4	21	47	62	79	121	172
2	15	34	44	57	87	124
1	11	25	33	42	65	93
1/0	9	22	28	36	56	79
2/0	8	18	23	30	46	65
3/0	6	15	20	25	38	55
4/0	5	12	16	20	32	45
250	4	10	13	16	25	35
300	3	8	11	14	22	31
350	3	7	9	12	19	27
400	3	6	8	11	17	24
500	2	5	7	9	14	20
600	1	4	6	7	11	16
700	1	4	5	6	10	14

THHN, THWN, THWN-2

Continued

Exhibit VII.62 Maximum Number of Same Size Compact Conductors in Type EB, PVC Conduit *Continued*

Compact Conductors

Type	Conductor Size (AWG/kcmil)	Metric Designator (Trade Size)						
		53 (2)	78 (3)	91 (3-1/2)	103 (4)	129 (5)	155 (6)
THHN, THWN, THWN-2	750	1	4	5	6	9	14
	1000	1	3	3	4	7	10
XHHW, XHHW-2	8	39	88	115	146	225	320
	6	29	65	85	109	167	238
	4	21	47	62	79	121	172
	2	15	34	44	57	87	124
	1	11	25	33	42	65	93
	1/0	9	22	28	36	56	79
	2/0	8	18	24	30	47	67
	3/0	6	15	20	25	38	55
	4/0	5	12	16	21	32	46
	250	4	10	13	17	26	37
	300	4	8	11	14	22	31
	350	3	7	10	12	19	28

400	3	7	9	11	17	25
500	2	5	7	9	14	20
600	1	4	6	7	11	16
700	1	4	5	6	10	14
750	1	3	5	6	9	13
1000	1	3	4	5	7	10

Note: Compact stranding is defined as the result of a manufacturing process where the standard conductor is compressed to the extent that the interstices (voids between strand wires) are virtually eliminated.

Source: NFPA 70, *National Electrical Code*®, NFPA, Quincy, MA, 2002 edition, Table C12(A).

Exhibit VII.63 Spacings for Vertical Conductor Supports

| | | Conductors | | | |
| | | Aluminum or Copper-Clad Aluminum | | Copper | |
Size of Wire	Support of Conductors in Vertical Raceways	m	ft	m	ft
18 AWG through 8 AWG	Not greater than	30	100	30	100
6 AWG through 1/0 AWG	Not greater than	60	200	30	100
2/0 AWG through 4/0 AWG	Not greater than	55	180	25	80
Over 4/0 AWG through 350 kcmil	Not greater than	41	135	18	60
Over 350 kcmil through 500 kcmil	Not greater than	36	120	15	50
Over 500 kcmil through 750 kcmil	Not greater than	28	95	12	40
Over 750 kcmil	Not greater than	26	85	11	35

Source: NFPA 70, *National Electrical Code®*, NFPA, Quincy, MA, 2002 edition, Table 300.19(A).

CONDUIT AND RACEWAY SYSTEMS

Exhibit VII.64 Supports for Rigid Metal Conduit (RMC)

Conduit Size		Maximum Distance between Rigid Metal Conduit Supports	
Metric Designator	Trade Size	m	ft
16–21	1/2–3/4	3.0	10
27	1	3.7	12
35–41	1-1/4–1-1/2	4.3	14
53–63	2–2-1/2	4.9	16
78 and larger	3 and larger	6.1	20

Source: NFPA 70, *National Electrical Code®*, NFPA, Quincy, MA, 2002 edition, Table 344.30(B)(2).

Exhibit VII.65 Supports for Rigid Nonmetallic Conduit (RNC)

Conduit Size		Maximum Spacing between Supports	
Metric Designator	Trade Size	mm or m	ft
16–27	1/2–1	900 mm	3
35–53	1-1/4–2	1.5 m	5
63–78	2-1/2–3	1.8 m	6
91–129	3-1/2–5	2.1 m	7
155	6	2.5 m	8

Source: NFPA 70, *National Electrical Code®*, NFPA, Quincy, MA, 2002 edition, Table 352.30(B).

Exhibit VII.66 Allowable Box Fill for Metal Boxes

Box Trade Size			Minimum Volume		Maximum Number of Conductors[a]						
mm	in.		cm³	in.³	18	16	14	12	10	8	6
100 × 32	(4 × 1-1/4)	round/octagonal	205	12.5	8	7	6	5	5	5	2
100 × 38	(4 × 1-1/2)	round/octagonal	254	15.5	10	8	7	6	6	5	3
100 × 54	(4 × 2-1/8)	round/octagonal	353	21.5	14	12	10	9	8	7	4
100 × 32	(4 × 1-1/4)	square	295	18.0	12	10	9	8	7	6	3
100 × 38	(4 × 1-1/2)	square	344	21.0	14	12	10	9	8	7	4
100 × 54	(4 × 2-1/8)	square	497	30.3	20	17	15	13	12	10	6
120 × 32	(4-11/16 × 1-1/4)	square	418	25.5	17	14	12	11	10	8	5
120 × 38	(4-11/16 × 1-1/2)	square	484	29.5	19	16	14	13	11	9	5
120 × 54	(4-11/16 × 2-1/8)	square	689	42.0	28	24	21	18	16	14	8
75 × 50 × 38	(3 × 2 × 1-1/4)	device	123	7.5	5	4	3	3	3	2	1
75 × 50 × 50	(3 × 2 × 2)	device	164	10.0	6	5	5	4	4	3	2
75 × 50 × 57	(3 × 2 × 2-1/4)	device	172	10.5	7	6	5	4	4	3	2
75 × 50 × 65	(3 × 2 × 2-1/2)	device	205	12.5	8	7	6	5	5	4	2
75 × 50 × 70	(3 × 2 × 2-3/4)	device	230	14.0	9	8	7	6	5	4	2
75 × 50 × 90	(3 × 2 × 3-1/2)	device	295	18.0	12	10	9	8	7	6	3

100 × 54 × 38	(4 × 2-1/8 × 1-1/2)	device	169	10.3	6	5	5	4	4	3	2
100 × 54 × 48	(4 × 2-1/8 × 1-7/8)	device	213	13.0	8	7	6	5	5	4	2
100 × 54 × 54	(4 × 2-1/8 × 2-1/8)	device	238	14.5	9	8	7	6	5	4	2
95 × 50 × 65	(3-3/4 × 2 × 2-1/2)	masonry box/gang	230	14.0	9	8	7	6	5	4	2
65 × 50 × 90	(3-3/4 × 2 × 3-1/2)	masonry box/gang	344	21.0	14	12	10	9	8	7	4
min. 44.5 depth	FS—single cover/gang (1-3/4)		221	13.5	9	7	6	6	5	4	2
min. 60.3 depth	FD—single cover/gang (2-3/8)		295	18.0	12	10	9	8	7	6	3
min. 44.5 depth	FS—multiple cover/gang (1-3/4)		295	18.0	12	10	9	8	7	6	3
min. 60.3 depth	FD—multipel cover/gang (2-3/8)		395	24.0	16	13	12	10	9	8	4

Where no volume allowances are required by 314.16(B)(2) through 314.16(B)(5) of the *NEC*.

Source: NFPA 70, *National Electrical Code®*, NFPA, Quincy, MA, 2002 edition, Table 314.16(A).

Exhibit VII.67 Volume Allowance Required per Conductor

Size of Conductor (AWG)	Free Space within Box for Each Conductor	
	cm³	in.³
18	24.6	1.50
16	28.7	1.75
14	32.8	2.00
12	36.9	2.25
10	41.0	2.50
8	49.2	3.00
6	81.9	5.00

Source: NFPA 70, *National Electrical Code®*, NFPA, Quincy, MA, 2002 edition, Table 314.16(B).

Exhibit VII.68 Minimum Burial Cover Requirements, 0 to 600 V, Nominal

Location of Wiring Method or Circuit	Column 1 Direct Burial Cables or Conductors		Column 2 Rigid Metal Conduit or Intermediate Metal Conduit		Column 3 Nonmetallic Raceways Listed for Direct Burial without Concrete Encasement or Other Approved Raceways	
	mm	in.	mm	in.	mm	in.
All locations not specified below	600	24	150	6	450	18
In trench below 50-mm (2-in.) thick concrete or equivalent	450	18	150	6	300	12
Under a building	0 (in raceway only)	0	0	0	0	0
Under minimum of 102-mm (4-in.) thick concrete exterior slab with no vehicular traffic and the slab extending not less than 152 mm (6 in.) beyond the underground installation	450	18	100	4	100	4

Continued

Exhibit VII.68 Minimum Burial Cover Requirements, 0 to 600 Volts, Nominal *Continued*

Location of Wiring Method or Circuit	Type of Wiring Method or Circuit					
	Column 1 Direct Burial Cables or Conductors		Column 2 Rigid Metal Conduit or Intermediate Metal Conduit		Column 3 Nonmetallic Raceways Listed for Direct Burial without Concrete Encasement or Other Approved Raceways	
	mm	in.	mm	in.	mm	in.
Under streets, highways, roads, alleys, driveways, and parking lots	600	24	600	24	600	24
One- and two-family dwelling driveways and outdoor parking areas, and used only for dwelling-related purposes	450	18	450	18	450	18
In or under airport runways, including adjacent areas where trespassing prohibited	450	18	450	18	450	18

Source: Modified from NFPA 70, *National Electrical Code®*, NFPA, Quincy, MA, 2002 edition, Table 300.5.

SECTION VIII

DOCUMENTATION AND DRAWINGS

This section provides reference materials to assist the engineer or designer in the preparation of fire alarm system drawings, whether in the office or in the field. It is also an invaluable reference for those who need to interpret drawings when reviewing, installing, or simply performing a field check of such systems. This information is critical for those who must interpret the notations of fire alarm system elements from the structural drawings of a building and to those who may need to verify the size and shape of structural elements and the type, quantity, and location of fire alarm system elements.

This section also illustrates standardized symbols for various components of building systems and miscellaneous components commonly used in the design of fire protection systems from several sources, including NFPA 170, *Standard for Fire Safety Symbols,* and NECA 100-1999, *Symbols for Electrical Construction Drawings.*

This section includes comprehensive lists from *NFPA 72®, National Fire Alarm Code®,* that itemize the information that must be provided for the system. These lists can be especially helpful when verifying that no essential information is missing from the working plans.

A fire alarm system record of completion is always completed after the system is initially installed and the installation wiring has been checked. The installing contractor provides a final copy after completion of the operational acceptance tests. When more than one contractor is responsible for the installation, each contractor is required to complete the portions of the form for which he or she has responsibility.

SYSTEM DOCUMENTATION

Exhibit VIII.1 System Documentation

Installation instructions (cut sheets) for every installed component

Battery calculation sheet (see Exhibit VIII.4)

Copies of UL certificates or FM placards

Copy of the original Inspection and Testing Form

Copy of the fire alarm system specification

Copy of the Fire Alarm System Record of Completion:

1. Name and address of the protected property
2. Owner and phone number of the protected property
3. Name and phone number of the Authority Having Jurisdiction
4. Type of fire alarm system or service
5. Applicable codes, standards, and other design criteria to which the system is required to comply
6. Type of building construction and occupancy
7. Fire department response point(s) and annunciator location(s)
8. Type(s) of fire alarm-initiating devices, supervisory alarm-initiating devices, and evacuation notification appliances to be provided
9. Calculations, for example, secondary supply and voltage drop calculations
10. Intended area(s) of coverage
11. Complete list of detection, evacuation signaling, and annunciator zones
12. Version of any applicable CAD software
13. Complete list of fire safety control functions
14. Complete sequence of operations detailing all inputs and outputs

Fire alarm system drawings:

1. Property name and location

Exhibit VIII.1 System Documentation *Continued*

---

2. Drawing developer's name, company name, address and telephone number
3. Revision number and date
4. Circuit identification
5. Terminal identification
6. Color-coding schemes
7. Location of devices and appliances
8. Type of devices, appliances, and equipment
9. Location of control equipment, annunciator(s), transmitters, and receivers
10. Location of control relays to other systems or suppression systems
11. Location of wiring and junction boxes
12. Type of cable or wiring method used for each circuit
13. Cable routing

An owner's manual:

1. Detailed narrative of the system including:
   a. Inputs
   b. Evacuation/relocation signaling
   c. Ancillary (control) functions
   d. Annunciation
   e. Sequence of operation, including an input/output matrix (see Exhibit VIII.5)
   f. Expansion capabilities
   g. Application limitations
   h. Revision of Basic Input Output Software (BIOS) installed

2. Operator instructions
   a. Basic system operation
   b. How to acknowledge an alarm
   c. How to silence alarm signals
   d. How to silence a trouble signal
   e. How to silence a supervisory signal

Continued

Exhibit VIII.1 System Documentation *Continued*

    f. How to reset the system

    g. Interpretation of system output and displays

    h. Operation of manual evacuation signaling

    i. Operation of ancillary controls

    j. How to perform routine maintenance, such as changing printer paper

3. Detailed description of testing and maintenance as required and recommended by the manufacturer under a maintenance contract for every component of the system.

    a. A list of system components that require periodic maintenance

    b. Step-by-step instructions for testing and maintenance for all installed equipment

    c. Schedule of testing and maintenance for all components per *NFPA 72®, National Fire Alarm Code®*, Chapter 7

    d. Detailed troubleshooting instructions for each type of anticipated trouble condition

    e. Service directory

Exhibit VIII.2 Fire Alarm System Record of Completion

Name of protected property: _____

Address: _____

Representative of protected property (name/phone): _____

Authority having jurisdiction: _____

Address/telephone number: _____

1. Type(s) of System or Service

_____ *NFPA 72®, National Fire Alarm Code®*, Chapter 3—Local

If alarm is transmitted to location(s) off premises, list where received: _____

_____ *NFPA 72*, Chapter 3—Emergency Voice/Alarm Service

Quantity of voice/alarm channels: _____ Single: _____ Multiple: _____

Continued

Exhibit VIII.2 Fire Alarm System Record of Completion *Continued*

Quantity of speakers installed: _____ Quantity of speaker zones: _____

Quantity of telephones or telephone jacks included in system: _____

_____ *NFPA 72*, Chapter 6—Auxillary

Indicate type of connection:

_____ Local energy _____ Shunt _____ Parallel telephone

Location of telephone number for receipt of signals: _____

_____ *NFPA 72*, Chapter 5—Remote Station

Alarm: _____

Supervisory: _____

_____ *NFPA 72*, Chapter 5—Proprietary

If alarms are retransmitted to public fire service communications centers or others, indicate
location and telephone numbers of the organization receiving alarm: _____

Indicate how alarm is retransmitted: _____

_____ NFPA 72, Chapter 5—Central Station

Prime contractor: _____

Central station location: _____

Means of transmission of signals from the protected premises to the central station:

_____ McCulloh   _____ Multiplex   _____ One-way radio

_____ Digital alarm communicator   _____ Two-way radio   _____ Others

Means of transmission of alarms to the public fire service communications center:

(a) _____

(b) _____

System location: _____

Continued

Exhibit VIII.2 Fire Alarm System Record of Completion *Continued*

	Organization name/phone	Representative name/phone
Installer		
Supplier		
Service organization		

Location of record (as-built) drawings: _____

Location of owners manuals: _____

Location of test reports: _____

A contract, dated _____, for test and inspection in accordance with NFPA standard(s)

No(s). _____, dated _____, is in effect.

2. Record of System Installation

(Fill out after installation is complete and wiring checked for opens, shorts, ground faults, and improper branching, but prior to conducting operational acceptance tests.)

This system has been installed in accordance with the NFPA standards as shown below, was inspected by _____ on _____, includes the devices shown below, and has been in service since _____.

_____ NFPA 72, Chapters   1   2   3   4   5   6   7   (circle all that apply)

_____ NFPA 70, *National Electrical Code®*, Article 760

_____ Manufacturer's instructions

_____ Other (specify):_____

Signed:_____ Date:_____ Organization:_____

3. Record of System Operation

All operational features and functions of this system were tested by _____ on _____,
and found to be operating properly in accordance with the requirements of:

_____ NFPA 72, Chapters   1   2   3   4   5   6   7   (circle all that apply)

_____ NFPA 70, *National Electrical Code®*, Article 760

_____ Manufacturer's instructions

_____ Other (specify):_____

Signed:_____ Date:_____ Organization:_____

Continued

Exhibit VIII.2 Fire Alarm System Record of Completion *Continued*

4. Alarm-Initiating Devices and Circuits

Quantity and class of initiating device circuits *(see NFPA 72, Table 3-5)*

Quantity: _____ Style: _____ Class: _____

MANUAL

(a) _____ Manual stations _____ Noncoded, activating _____ Transmitters _____ Coded

(b) _____ Combination manual fire alarm and guard's tour coded stations

AUTOMATIC

Coverage: Complete: _____ Partial: _____

(a) _____ Smoke detectors _____ Ion _____ Photo

(b) _____ Duct detectors _____ Ion _____ Photo

(c) _____ Heat detectors _____ FT _____ RR _____ FT/RR _____ RC

Continued

(d) _____ Sprinkler waterflow switches:

_____ Transmitters _____ Noncoded, activating _____ Coded

(e) _____ Other (list):

5. **Supervisory Signal-Initiating Devices and Circuits** (use blanks to indicate quantity of devices)

GUARD'S TOUR

(a) _____ Coded stations

(b) _____ Noncoded stations, activating _____ transmitters

(c) _____ Compulsory guard tour system comprised of _____ transmitter stations and _____ intermediate stations

Note: Combination devices are recorded under 4(b) and 5(a).

SPRINKLER SYSTEM

(a) _____ Coded valve supervisory signaling attachments

_____ Value supervisory switches, activating _____ transmitters

(b) _____ Building temperature points

Exhibit VIII.2 Fire Alarm System Record of Completion *Continued*

(c) ——— Site water temperature points

(d) ——— Site water supply level points

Electric fire pump:

(e) ——— Fire pump power

(f) ——— Fire pump running

(g) ——— Phase reversal

Engine-driven fire pump:

(h) ——— Selector in auto position

(i) ——— Engine or control panel trouble

(j) ——— Fire pump running

Engine-driven generator:

(k) ——— Selector in auto position

(l) ——— Control panel trouble

(m) _____ Transfer switches

(n) _____ Engine running

Other supervisory function(s) (specify): _____

_____

_____

_____

6. Alarm Notification Appliances and Circuits

Quantity and class (see NFPA 72, Table 3-7) of notification appliance circuits connected to the system:

Types and quantities of notification appliances installed: Quantity: _____ Style: _____ Class: _____

(a) _____ Bells _____ Inch

(b) _____ Speakers

(c) _____ Horns

(d) _____ Chimes

(e) _____ Other: _____

_____

_____

_____

Continued

Exhibit VIII.2 Fire Alarm System Record of Completion *Continued*

(f) _____ Visual signals   Type: _____

_____ with audible _____ w/o audible

(g) _____ Local annunciator

7. Signaling Line Circuits

Quantity and class (*see NFPA 72, Table 3-6*) of signaling line circuits connected to system:

Quantity: _____   Style: _____   Class: _____

8. System Power Supplies

(a) Primary (main): _____   Nominal voltage: _____   Current rating: _____

Overcurrent protection: Type: _____   Current rating: _____

Location: _____

Continued

(b) Secondary (standby):

——— Storage battery: Amp-hour rating:

——— Calculated capacity to drive system, in hours: ——— 24 ——— 60

——— Engine-driven generator dedicated to fire alarm system:

Location of fuel storage:

(c) Emergency or standby system used as backup to primary power supply, instead of using a secondary power supply:

——— Emergency system described in NFPA 70, Article 700

——— Legally required standby system described in NFPA 70, Article 701

——— Optional standby system described in NFPA 70, Article 702, which also meets the performance requirements of Article 700 or 701

Exhibit VIII.2 Fire Alarm System Record of Completion *Continued*

9. System Software

(a) Operating system software revision level(s): _____

(b) Application software revision level(s): _____

(c) Revision completed by: _____ _____
                                              (name)   (firm)

10. Comments: _____

(signed) for central station or alarm service company or installation contractor/supplier   (title)   (date)

Frequency of routine tests and inspections, if other than in accordance with the referenced NFPA standard(s): _____

_____

_____

_____

System deviations from the referenced NFPA standard(s) are: _____

_____

_____

_____

(signed) for central station or alarm service company or installation contractor/supplier   (title)   (date)

Upon completion of the system(s) satisfactory test(s) witnessed (if required by the authority having jurisdiction):

(signed) representative of the authority having jurisdiction   (title)   (date)

Source: NFPA 72®, National Fire Alarm Code®, NFPA, Quincy, MA, 1999 edition, Fig. 1-6.2.1.

Exhibit VIII.3 Inspection and Testing Form

INSPECTION AND TESTING FORM

DATE: _____

TIME: _____

PROPERTY NAME (USER)

Name: _____

Address: _____

Owner Contact: _____

Telephone: _____

SERVICE ORGANIZATION

Name: _____

Address: _____

Representative: _____

License No.: _____

Telephone: _____

MONITORING ENTITY

Contact: _____

Telephone: _____

Monitoring: _____

Account Ref. No.: _____

TYPE TRANSMISSION

☐ McCulloh

☐ Multiplex

☐ Digital

☐ Reverse Priority

☐ RF

☐ Other (Specify) _____

Control Unit Manufacturer: _____

Circuit Styles: _____

APPROVING AGENCY

Contact: _____

Telephone: _____

SERVICE

☐ Weekly

☐ Monthly

☐ Quarterly

☐ Semiannually

☐ Annually

☐ Other (Specify) _____

Model No.: _____

Continued

Exhibit VIII.3 Inspection and Testing Form *Continued*

Number of Circuits: _____

Software Rev.: _____

Last Date System Had Any Service Performed: _____

Last Date That Any Software or Configuration Was Revised: _____

ALARM-INITIATING DEVICES AND CIRCUIT INFORMATION

Quantity	Circuit Style	
_____	_____	Manual Fire Alarm Boxes
_____	_____	Ion Detectors
_____	_____	Photo Detectors
_____	_____	Duct Detectors
_____	_____	Heat Detectors
_____	_____	Waterflow Switches
_____	_____	Supervisory Switches
_____	_____	Other (Specify): _____

ALARM NOTIFICATION APPLIANCES AND CIRCUIT INFORMATION

Quantity	Circuit Style	
_____	_____	Bells
_____	_____	Horns
_____	_____	Chimes
_____	_____	Strobes
_____	_____	Speakers
_____	_____	Other (Specify): _____

No. of alarm notification appliance circuits: _____
Are circuits monitored for integrity?   ☐ Yes   ☐ No

SUPERVISORY SIGNAL-INITIATING DEVICES AND CIRCUIT INFORMATION

Quantity	Circuit Style	
_____	_____	Building Temp.
_____	_____	Site Water Temp.

Continued

Exhibit VIII.3 Inspection and Testing Form *Continued*

_____	Site Water Level
_____	Fire Pump Power
_____	Fire Pump Running
_____	Fire Pump Auto Position
_____	Fire Pump or Pump Controller Trouble
_____	Fire Pump Running
_____	Generator In Auto Position
_____	Generator or Controller Trouble
_____	Switch Transfer
_____	Generator Engine Running
_____	Other: _____

SIGNALING LINE CIRCUITS

Quantity and style (See *NFPA 72*, Table 3-6) of signaling line circuits connected to system:

Quantity _____ Style(s) _____

SYSTEM POWER SUPPLIES

a. Primary (Main): Nominal Voltage _____ , Amps _____

Overcurrent Protection: Type _____ , Amps _____

Continued

Location (of Primary Supply Panelboard): _____

Disconnecting Means Location: _____

b. Secondary (Standby): _____

Storage Battery: _____ Amp-Hr. Rating

Calculated capacity to operate system, in hours: _____ 24 _____ 60

_____ Engine-driven generator dedicated to fire alarm system:

Location of fuel storage: _____

TYPE BATTERY

☐ Dry Cell   ☐ Nickel Cadmium   ☐ Sealed Lead Acid

☐ Lead Acid   ☐ Other (Specify): _____

c. Emergency or standby system used as a backup to primary power supply, instead of
using a secondary power supply:

_____ Emergency system described in NFPA 70, Article 700

_____ Legally required standby described in NFPA 70, Article 701

_____ Optional standby system described in NFPA 70, Article 702, which also
meets the performance requirementsof Article 700 or 701.

Exhibit VIII.3 Inspection and Testing Form *Continued*

PRIOR TO ANY TESTING

NOTIFICATIONS ARE MADE	Yes	No	Who	Time
Monitoring Entity	☐	☐		
Building Occupants	☐	☐		
Building Management	☐	☐		
Other (Specify)	☐	☐		
AHJ (Notified) of Any Impairments	☐	☐		

SYSTEM TESTS AND INSPECTIONS

TYPE	Visible	Functional	Comments
Control Unit	☐	☐	
Interface Eq.	☐	☐	
Lamps/LEDS	☐	☐	
Fuses	☐	☐	
Primary Power Supply	☐	☐	
Trouble Signals	☐	☐	
Disconnect Switches	☐	☐	
Ground-Fault Monitoring	☐	☐	

	Visible	Functional	Comments
SECONDARY POWER			
TYPE			
Battery Condition	☐	☐	
Load Voltage		☐	
Discharge Test		☐	
Charger Test		☐	
Specific Gravity			
TRANSIENT SUPPRESSORS	☐	☐	
REMOTE ANNUNCIATORS	☐		
NOTIFICATION APPLIANCES			
Audible	☐	☐	
Visual	☐	☐	
Speakers	☐	☐	
Voice Clarity		☐	

Continued

Exhibit VIII.3 Inspection and Testing Form *Continued*

INITIATING AND SUPERVISORY DEVICE TESTS AND INSPECTIONS

Loc. & S/N	Device Type	Visual Check	Functional Test	Factory Setting	Meas. Setting	Pass	Fail
		☐	☐			☐	☐
		☐	☐			☐	☐
		☐	☐			☐	☐
		☐	☐			☐	☐
		☐	☐			☐	☐
		☐	☐			☐	☐

Comments: _____

Continued

EMERGENCY COMMUNICATIONS EQUIPMENT

	Visual	Functional	Comments
Phone Set	☐	☐	_____
Phone Jacks	☐	☐	_____
Off-Hook Indicator	☐	☐	_____
Amplifier(s)	☐	☐	_____
Tone Generator(s)	☐	☐	_____
Call-in Signal	☐	☐	_____
System Performance	☐	☐	_____

INTERFACE EQUIPMENT

	Visual	Device Operation	Simulated Operation
(Specify) _____	☐	☐	☐
(Specify) _____	☐	☐	☐
(Specify) _____	☐	☐	☐

SPECIAL HAZARD SYSTEMS

	Visual	Device Operation	Simulated Operation
(Specify) _____	☐	☐	☐
(Specify) _____	☐	☐	☐
(Specify) _____	☐	☐	☐

Exhibit VIII.3 Inspection and Testing Form *Continued*

Special Procedures: _____

Comments: _____

SUPERVISING STATION MONITORING	Yes	No	Time	Comments
Alarm Signal	☐	☐		
Alarm Restoration	☐	☐		
Trouble Signal	☐	☐		
Supervisory Signal	☐	☐		
Supervisory Restoration	☐	☐		

NOTIFICATIONS THAT TESTING IS COMPLETE

	Yes	No	Who	Time
Building Management	☐	☐	_____	_____
Monitoring Agency	☐	☐	_____	_____
Building Occupants	☐	☐	_____	_____
Other (Specify)	☐	☐	_____	_____

The following did not operate correctly: _____

_____

System restored to normal operation: Date: _____ Time: _____

THIS TESTING WAS PERFORMED IN ACCORDANCE WITH APPLICABLE NFPA STANDARDS.

Name of Inspector: _____ Date: _____ Time: _____

Signature: _____

Name of Owner or Representative: _____

Date: _____ Time: _____

Signature: _____

Source: NFPA 72®, National Fire Alarm Code®, NFPA, Quincy, MA, 1999 edition, Fig. 7-5.2.2.

Exhibit VIII.4 Example Battery Calculation Sheet

Item	Description	Standby Current per Unit (Amps)		Qty		Total System Standby Current per Item (Amps)	Alarm Current per Unit (Amps)		Qty		Total System Alarm Current per Item (Amps)
A	FACP	0.1200	×	1	=	0.1200	1.5000	×	1	=	1.5000
B	Smoke detectors	0.0005	×	15	=	0.0075	0.0010	×	5	=	0.0050
C	Relays	0.0070	×	4	=	0.0280	None	×	4	=	None
D	Horn/strobes	None	×	23	=	None	0.0780	×	23	=	1.7940
			×		=			×		=	
			×		=			×		=	
			×		=			×		=	
			×		=			×		=	
		Total System Standby Current (Amps)				0.1555	Total System Alarm Current (Amps)				3.2990

Maximum Quiescent Load (Standby): 24 hr   Total Alarm Load: 5 min × 1/60 = 0.083 hr

Required Standby Time (hr)		Total System Standby Current (Amps)		Required Standby Capacity (Amp-hr)
24	×	0.1555	=	3.7320

Required Alarm Time (hr)		Total System Alarm Current (Amps)		Required Alarm Capacity (Amp-hr)
0.083	×	3.2990	=	0.2738

Required Standby Capacity (Amp-hr)		Required Alarm Capacity (Amp-hr)		Total Required Capacity (AMP-hr)
3.7320	+	0.2738	=	4.0058

Total Required Capacity (AMP-hr)		Optional Factor of Safety		Adjusted Battery Capacity (Amp-hr)
4.0058	×	1.2	=	4.8069

Exhibit VIII.5 Typical Fire Alarm System Input/Output Matrix

Inputs	Actuate Evacuation Signals	Actuate Supervisory Signal	Initiate Fan Shutdown	Recall Elevators	Transmit Signal to CS	Comments
Manual Fire Alarm Boxes	×				×	
Smoke Detectors	×				×	
Waterflow	×				×	
Sprinkler Control Valve		×			×	
Fire Pump Running		×			×	
Fire Pump Phase Reversed		×			×	
In-Duct Smoke Detectors		×	×		×	
Elevator Lobby Smoke Detectors	×			×	×	

ELECTRICAL DRAWING SYMBOLS

Exhibit VIII.6 Raceway and Box Symbols

Symbol	Description
———————	Conduit concealed in construction in finished areas, exposed in unfinished areas
- - - - - - -	Conduit concealed in or under floor slab
∿	Nonrigid raceway system
—— NE ——	Normal/emergency circuit
—— EB ——	Emergency battery system wiring, minimum: #10 AWG
- - - P - - -	Underfloor power raceway
- - - T - - -	Underfloor telecommunications raceway
- - - PT - - -	Underfloor raceway for power and telecommunications
- - - S - - -	Underfloor signal raceway
- - - PTD - - -	Underfloor raceway for power, telephone, and data
- - - UCP - - -	Undercarpet flat conductor cable (FCC) wiring system, power
- - - UCT - - -	Undercarpet flat conductor cable (FCC) wiring system, telephone
- - - UCD - - -	Undercarpet flat conductor cable (FCC) wiring system, data
——————⌐	Conduit stub. Terminate with bushing or cap if underground
———————O	Conduit turning up
———————●	Conduit turning down
SZ 2 C. 4 #1 & 1 #6 GND. or SZ 53 mm 4 #1 & 1 #6 GND	Indicates trade size 2" or 53-mm conduit with (4) #1 AWG and (1) #6 AWG ground

Continued

ELECTRICAL DRAWING SYMBOLS

Exhibit VIII.6 Raceway and Box Symbols *Continued*

Symbol	Description
(2) SZ 2 C. 4 #1 & 1 #6 GND. or (2) SZ 53 mm 4 #1 & 1 #6 GND.	Indicates (2) trade size 2" or 53-mm conduits with (4) #1 AWG and (1) #6 AWG ground conductors in each conduit
L211–1.3 ⫴⟶⟶	Homerun to panelboard; number of arrows indicates number of circuits (example: homerun to panel L211 ckts. #1 and #3)
∿•	Flexible connection to equipment
—•	Direct connection to equipment
P ■ ₂	Power pole with devices indicated in the specifications and on the drawings, "P" indicates type, "2" indicates circuit
T ■	Telecom pole with devices indicated in the specifications and on the drawings, "T" indicates type
TP ■ ₂	Telecom/power pole with devices indicated in the specifications and on the drawings, "TP" indicates type, "2" indicates power circuit
PB or ▨	Pull box—size as indicated or required
TR TR TR	Cabletray size as indicated
TR TR TR	Cabletray size as indicated, concealed
way way way	Wireway, size as indicated or required

Source: NECA 100-1999, *Symbols for Electrical Construction Drawings* (ANSI), National Electrical Contractors Association, Bethesda, MD, 1999.

ELECTRICAL DRAWING SYMBOLS

Exhibit VIII.7 Lighting Fixture Symbols

Symbol	Description
○ □ △	Luminaire (drawn to approximate shape and to scale or large enough for clarity)
	Luminaire strip type (length drawn to scale)
	Fluorescent strip lighting
	Fixture—double or single head spotlight
	Exit lighting fixture, arrows, and exit face as indicated on DWGS (mounting heights to be determined by job specifications)
▽ ▽ ▽	Light track, length as indicated on the drawings, with number of fixtures as indicated on drawings, and as indicated in the fixture schedule
	Emergency battery remote lighting heads
	Emergency battery unit with lighting heads
○	Surface mounted fixture
∅	Recessed fixture
♀	Wall mounted fixture
⊙	Suspended, pendant, chain, stem or cable hung fixture
◑	Luminaire providing emergency illumination (filled in)

Source: NECA 100-1999, *Symbols for Electrical Construction Drawings* (ANSI), National Electrical Contractors Association, Bethesda, MD, 1999.

ELECTRICAL DRAWING SYMBOLS

Exhibit VIII.8 Outlet and Receptacle Symbols

Symbol	Description
⊡ F	Floor duplex receptacle. F = flush MTD. S = surface MTD.
⊖	Duplex convenience receptacle; 20A 125V; wall mounted device, at 18" AFF. U.O.I.
EP-2 CKT.1 ⬤=	Duplex convenience receptacle on emergency/standby circuit; specify panelboard and circuit
⊖	Single convenience receptacle
EP-2 CKT.3 ⬤=	Single convenience receptacle on emergency/standby circuit; specify panelboard and circuit
⬦	Double duplex convenience receptacle
EP-2 CKT.5 ⬥=	Double duplex convenience receptacle on emergency/standby circuit; specify panelboard and circuit
A ▬▪▪▪▪▪▬	Multi-outlet assembly with outlets on centers as indicated on the drawings and in the specifications, mounted 6 in. above counter or at height as directed, A—indicates type
⏀ ⏀1	Multi-outlet assembly, devices as indicated
⏀1 or 1	Special receptacle—typical notation: 1—indicates example "1" = __A, ___/___V., __ Pole, __ Wire, __ NEMA __-__ "2" = __A,___/___V., __ Pole, __ Wire, __ NEMA __-__ "3" = __A,___/___V., __ Pole, __ Wire, __ NEMA __-__
©—	Clock hanger outlet recessed mounted 8'-0" AFF or 8" below ceiling as appropriate and as directed

Exhibit VIII.8 Outlet and Receptacle Symbols *Continued*

Symbol	Description
▼ ⏀ F	Flush mounted floor box, adjustable, with both power and voice/data receptacles
J ⬜ J A×B×C	Junction box, "A × B × C" indicates dimensions of junction box in either inches or centimeters

Source: NECA 100-1999, *Symbols for Electrical Construction Drawings* (ANSI), National Electrical Contractors Association, Bethesda, MD, 1999.

Exhibit VIII.9 Modifiers for Receptacles and Outlets

Typical Additional Notations

"a"	=	Switched outlet, "a"—indicates switch control
"B"	=	Pedestal mounted on bench top
"BF"	=	Below floor
"C"	=	Mounted 6" above counter of 42" AFF; coordinate exact mounting height with architectural drawings
"CLG"	=	Ceiling mounted
"D"	=	Dedicated device on individual branch circuit
"E"	=	Emergency
"EXIST."	=	Existing device/equipment
"F"	=	Flush floor box with fire/smoke rated penetration
"GFCI"	=	Ground fault circuit interupter, personal protection
"GFPE"	=	Ground fault protection of equipment
"H"	=	Horizontally mounted
"IG"	=	Isolated ground receptacle with separate green ground conductor to isolated ground bus in panel
"M"	=	Modular furniture service—provide flexible connection, coordinate exact location with furniture plans
"PED"	=	Pedestal mounted with two-hour fire/smoke rated penetration
"PT"	=	Poke thru with two-hour fire/smoke rated penetration
"S"	=	Surface mounted floor box
"SP"	=	Surge protection receptacle
"T"	=	Tamper resistant safety receptacle
"TL"	=	Twist-lock
"W"	=	Wall mounted device at 48" AFF unless otherwise indicated
"WP"	=	Weatherproof receptacle with "NRTL" listed coverplate for wet location with plug installed; MTD; 48" AFF unless otherwise indicated
"XX"	=	Dimensioned height

Source: NECA 100-1999, *Symbols for Electrical Construction Drawings* (ANSI), National Electrical Contractors Association, Bethesda, MD, 1999.

ELECTRICAL DRAWING SYMBOLS

Exhibit VIII.10 Switch and Sensor Symbols

Symbol	Description
S	Single pole switch
S_2	Double pole switch
S_3	Three-way switch
S_4	Four-way switch
S_a	Switch control (lowercase letter)
S_{CB}	Circuit breaker switch
S_{DT}	Single pole/double throw switch
S_G	Glow switch toggle, glows in off position
S_H	Horizontally mounted—with on position to the left
S_K	Key operated switch
S_{KP}	Key operated switch with pilot light on when switch is on
S_{LV}	Low voltage switch
S_{LM}	Low voltage master switch
S_{MC}	Momentary contact switch
S_P	Switch with pilot light on when switch is on
S_T	Timer switch
S_{WP}	Weatherproof single pole switch
$\boxed{D}$	Dimmer switch. Rated 1000W, unless otherwise indicated. "LV" = low voltage "FL" = fluorescent.
$\langle M \rangle$	Occupancy sensor, wall mounted with off—auto override switch.
$\langle M \rangle_P$	Occupancy sensor—ceiling mounted, "P"—indicates multiple switches wired in parallel.

Source: NECA 100-1999, *Symbols for Electrical Construction Drawings* (ANSI), National Electrical Contractors Association, Bethesda, MD, 1999.

ELECTRICAL DRAWING SYMBOLS

Exhibit VIII.11 HVAC Equipment and Control Symbols

Symbol	Description
EP	Electric/pneumatic switch
HOA	Hand/off/automatic selector switch
FS	Flow switch
LS	Limit switch
PE	Pneumatic/electric switch
PC	Photo cell or photo control
PS	Pressure switch
HVAC	HVAC Control Panel
③	Motor "3"—indicates horsepower
Ⓓ	Motorized damper
Ⓞ	Ceiling fan
✕	Paddle fan
⊠	Wall fan

Source: NECA 100-1999, *Symbols for Electrical Construction Drawings* (ANSI), National Electrical Contractors Association, Bethesda, MD, 1999.

ELECTRICAL DRAWING SYMBOLS

Exhibit VIII.12 Security Symbols

Symbol	Description
C ◁ WP	CCVT camera. "WP" indicates weatherproof exterior camera
CCTV	CCTV Coaxial cable outlet and power outlet
MTV	CCTV monitor outlet
B ⊃	Doorbell
B /	Door buzzer
B ⊨	Door chime
DR	Electric door opener
ES	Electric door strike
IC	Intercom unit—flush MTD
MI	Master intercom and directory unit
MD	Motion detector
ML	Security door alarm magnetic lock
CR WP	Security card reader. "WP" indicates weatherproof
SCP	Security control panel
DC	Security door contacts
▪	Security exit push button
K	Security keypad

Source: NECA 100-1999, *Symbols for Electrical Construction Drawings* (ANSI), National Electrical Contractors Association, Bethesda, MD, 1999.

ELECTRICAL DRAWING SYMBOLS

Exhibit VIII.13 Communication Symbols

Symbol	Description
▼	Telephone outlet
▼_W	Telephone outlet—wall mounted
▼_F	Telephone outlet floor type. "F" indicates flush mounted. "S" indicates surface mounted
▭	Equipment cabinet
▄	Equipment rack—wall mounted
▦	Equipment rack—free standing
TCC	Terminal cabinet with ¾" plywood backing
▨	Plywood backboard

Source: NECA 100-1999, *Symbols for Electrical Construction Drawings* (ANSI), National Electrical Contractors Association, Bethesda, MD, 1999.

ELECTRICAL DRAWING SYMBOLS

Exhibit VIII.14 Site Work Symbols

Symbol	Description
– – – – – UF – – – – –	Underground feeder
– – – – – UT – – – – –	Underground telephone
– – – – UFA – – – –	Underground fire alarm
– – – – UTV – – – –	Underground television (CATV)
——— E ———	Above ground pole mounted electrical
——— T ———	Above ground pole mounted telephone
——— F ———	Above ground pole mounted fire alarm
——— TV ———	Above ground pole mounted television (CATV)
—[MH]—	Manhole
—[HH]—	Handhole
Ⓟ Pxxxx	Utility pole; "Pxxxx" indicates pole number

Source: NECA 100-1999, *Symbols for Electrical Construction Drawings* (ANSI), National Electrical Contractors Association, Bethesda, MD, 1999.

ELECTRICAL DRAWING SYMBOLS

Exhibit VIII.15 Abbreviations

Abbreviation	Description
1P	One pole
2P	Two pole
3P	Three pole
4P	Four pole
1P1W	One pole, one wire
1P2W	One pole, two wire
2P2W	Two pole, two wire
2P3W	Two pole, three wire
3P2W	Three pole, two wire
3P3W	Three pole, three wire
3P4W	Three pole, four wire
4P4W	Four pole, four wire
A	Ampere
AC	Alternating current
AF	AMP frame
AFF	Above finished floor
AFG	Above finished grade
AIC	Ampere interrupting capacity
AL	Aluminum
AS	AMP switch
AT	AMP trip
ARCH	Architect
ATS	Automatic transfer switch
AUD	Audiometer box connection
A/V	Audio visual
BLDG	Building
C	Conduit (generic term for raceway); provide as specified
CAM	Camera
CAT	Catalog
CATV	Cable television
CB	Circuit breaker
CKT	Circuit
COL	Column

Exhibit VIII.15 Abbreviations *Continued*

Abbreviation	Description
C.T.	Current transformer
CU	Copper
℄	Centerline
DC	Direct Current
Δ	Delta
DISC	Disconnect
DWG	Drawing
DT	Dusttight[a]
E	Wired on emergency circuit
EC	Electrical contractor
EMT	Electric metallic tubing
EOL	End of line
EWC	Electric water cooler
EXIST.	Existing
F	Flush
FA	Fire alarm
FBO	Furnished by others
FC	Fire protection contractor
FDN	Foundation
FLA	Full load amps
FMC	Flexible metallic conduit
FRE	Fiberglass reinforced epoxy conduit
GC	General contractor
GFCI	Ground fault circuit interrupter
GFPE	Ground fault protection equipment
GND	Grounded
GRC	Galvanized rigid conduit
HP	Horsepower
HV	High voltage
HVAC	Heating, ventilating and air conditioning
Hz	Hertz (cycle per second)
IAM	Individual addressable module
IG	Isolated ground

Continued

Exhibit VIII.15 Abbreviations *Continued*

Abbreviation	Description
IMC	Intermediate metal conduit
JB	Junction box
KCMIL	Thousand circular mils
K/O	Knock-out
KVA	Kilovolt ampere
KVAR	Kilovolt ampere reactive
KW	Kilowatt
LFMC	Liquidtight flexible metallic conduit
LFNC	Liquidtight flexible nonmetallic conduit
LP	Lighting panelboard
LS	Limit switch
LTG	Lighting
LV	Low voltage
MC	Metal clad cable
MCB	Main circuit breaker
MCC	Motor control center
MDP	Main distribution panel
MISC	Miscellaneous
MLO	Main lugs only
MOD	Motor operated disconnect switch
MTD	Mounted
MTG	Mounting
MTS	Manual transfer switch
N/A	Not applicable
NC	Normally closed
NEC	National Electrical Code
NIC	Not in contract
NL	Night light
NM	Nonmetallic sheathed cable
NO	Normally open
NRTL	Nationally recognized testing lab
#	Number
NTS	Not to scale

Exhibit VIII.15 Abbreviations *Continued*

Abbreviation	Description
P	Pole
PB	Pull box
PC	Plumbing system contractor
PH φ	Phase
PNL	Panel(board)
PIV	Post indicating valve
PP	Power panel
PR	Pair
PRI	Primary
PT	Potential transformer
PVC	Polyvinyl chloride conduit
PWR	Power
REC	Recessed
RT	Raintight[a]
RSC	Rigid steel conduit
S	Surface mounted
SEC	Secondary
SIG	Signal
SN	Solid neutral
SP	Spare
SPL	Splice
SS	Stainless steel
STP	Shielded twisted pair
STL	Carbon steel
SUSP	Suspended
SW	Switch
SWBD	Switchboard
TC	Telephone cabinet
TCI	Telecommunications cabling installer
TEL/DATA	Telephone/data
TEL	Telephone
TERM	Terminal(s)
TYP	Typical

Continued

ELECTRICAL DRAWING SYMBOLS

Exhibit VIII.15 Abbreviations *Continued*

Abbreviation	Description
UG	Underground
UTP	Unshielded twisted pair
V	Volt
VT	Vaportight[a]
U.O.I.	Unless otherwise indicated
Y	Wye
W	Watt
WH	Watthour
WP	Weatherproof
WT	Watertight[a]
XFMR	Transformer
XP	Explosion proof[a]
ZAM	Zone adapter module
+72	Mounting units to centerline above finished floor or grade

[a]It is recommended that the appropriate NEMA designation be used in place of this abbreviation.

Source: NECA 100-1999, *Symbols for Electrical Construction Drawings* (ANSI), National Electrical Contractors Association, Bethesda, MD, 1999.

ALARM AND RELATED SYMBOLS

Exhibit VIII.16 Fire Alarm Communication and Control Symbols

Symbol	Description
M (master box symbol)	Fire alarm master box
(fire fighter's phone symbol)	Fire fighter's phone
FT	Coded transmitter
FTR	Fire alarm transponder or transmitter[a]
ESR	Elevator status/recall[a]
DK	Drill key switch
K (key repository symbol)	Key repository (knox box)
FAA	Annunciator panel
FACP	Fire alarm control panel
EVAC	Voice evacuation panel
FATC	Fire alarm terminal cabinet
BATT	Battery pack and charger
ASFP	Air sampling control/detector panel with associated air sampling piping network
FAC	Fire alarm communicator[a]
IAM	Individual addressable module
ZAM	Zone adapter module

[a]NFPA symbol. © National Fire Protection Association.

Source: NECA 100-1999, *Symbols for Electrical Construction Drawings* (ANSI), National Electrical Contractors Association, Bethesda, MD, 1999.

ALARM AND RELATED SYMBOLS

Exhibit VIII.17 Fire Alarm Initiating Device Symbols

Symbol	Description
F	Manual fire alarm box
HL	Manual fire alarm box–Halon[a]
CO2	Manual fire alarm box–Carbon dioxide[a]
DC	Manual fire alarm box–Dry chemical[a]
FO	Manual fire alarm box–Foam[a]
WC	Manual fire alarm box–Wet chemical[a]
CA	Manual fire alarm box–Clean agent[a]
WM	Manual fire alarm box–Water mist[a]
DL	Manual fire alarm box–Deluge sprinkler[a]
RTS	Remote station for duct mounted smoke detectors
R/F	Combination—Rate of rise and fixed temperature heat detector[a]

**Exhibit VIII.17 Fire Alarm
Initiating Device Symbols** *Continued*

Symbol	Description
R/C	Rate compensation heat detector[a]
F	Fixed temperature heat detector[a]
R	Rate of rise only heat detector[a]
→	Line-type (heat-sensitive cable) heat detector[a]
P	Photoelectric detector[a]
I	Ionization products of combustion detector[a]
BT	Beam transmitter[a]
BR	Beam receiver[a]
ASD	Air sampling[a]
	Smoke detector for duct[a]

Continued

**Exhibit VIII.17 Fire Alarm
Initiating Device Symbols** *Continued*

Symbol	Description
▲ (triangle in filled circle)	Gas detector[a]
∧ IR	Infrared flame detector[a]
∧ UV	Ultraviolet flame detector[a]
∧ UV/IR	Ultraviolet/infrared flame detector[a]
Ⓓ FS	Motor operated fire/smoke duct damper
FS	Water flow switch
PS	Pressure switch
TS	Tamper switch
Flow detector/switch symbol	Flow detector/switch[a]
Pressure detector/switch symbol	Pressure detector/switch[a]

ALARM AND RELATED SYMBOLS

Exhibit VIII.17 Fire Alarm
Initiating Device Symbols *Continued*

Symbol	Description
	Level detector/switch[a]
	Tamper switch[a]
	Valve with tamper switch[a]
PIV	Post indicator valve
EOL	End-of-line resistor

[a]NFPA symbol. © National Fire Protection Association.

Source: NECA 100-1999, Symbols for Electrical Construction Drawings (ANSI), National Electrical Contractors Association, Bethesda, MD, 1999.

Exhibit VIII.18 Fire Alarm Notification Appliance and Control Symbols

Symbol	Description
(CR)	Control relay
DH	Door holder
F ◇	Mini-horn and strobe
H ◀	Horn unit only
M ◁	Mini-horn[a]
S ◀	Strobe unit only
☼ ◁	Horn with strobe as separate assembly[a]
☼◁	Horn with strobe as one assembly[a]
F ○	Bell and strobe
F /	Buzzer and strobe
F =	Chime and strobe
F (S)	Speaker and strobe

ALARM AND RELATED SYMBOLS

Exhibit VIII.18 Fire Alarm Notification Appliance and Control Symbols *Continued*

Symbol	Description
	LED pilot light
WP	Red indicating beacon; "WP" indicates weatherproof
	Bell (gong)[a]
	Water motor alarm (water motor gong)[a]

[a]NFPA symbol. © National Fire Protection Association.

Source: NECA 100-1999, *Symbols for Electrical Construction Drawings* (ANSI), National Electrical Contractors Association, Bethesda, MD, 1999.

STANDARD FIRE SAFETY SYMBOLS

Exhibit VIII.19 Wall and Parapet Symbols

Symbol	Description
	Wall
	Smoke barrier
	1/2-hr fire-rated
	1/2-hr fire-rated/smoke barrier
	3/4-hr fire-rated
	3/4-hr fire-rated/smoke barrier
	1-hr fire-rated
	1-hr fire-rated/smoke barrier
	2-hr fire-rated
	2-hr fire-rated/smoke barrier
	3-hr fire-rated
	3-hr fire-rated/smoke barrier
	4-hr fire-rated
	4-hr fire-rated/smoke barrier

Source: NFPA 170, *Standard for Fire Safety Symbols*, NFPA, Quincy, MA, 1999 edition.

STANDARD FIRE SAFETY SYMBOLS

Exhibit VIII.20 Floor, Wall and Roof Opening Symbols and Their Protection

Symbol	Description
	Opening in wall
	Rated fire door in wall (less than 3 hr)
	Fire door in wall (3-hr rated)
	Elevator in combustible shaft
	Elevator in noncombustible shaft
	Open hoistway
	Escalator
	Stairs in combustible shaft
	Stairs in fire-rated shaft
	Stairs in open shaft
	Skylight

Source: NFPA 170, *Standard for Fire Safety Symbols*, NFPA, Quincy, MA, 1999 edition.

STANDARD FIRE SAFETY SYMBOLS

Exhibit VIII.21 Special Symbols for Cross Sections

Symbol	Description
⊓⊓⊓⊓⊓⊓⊓⊓	Wood joisted floor or roof
(Steel deck on steel joists)	Other floors or roofs
══════════	Floor/ceiling or roof/ceiling assembly
≡⦀≡⦀≡⦀≡⦀	Floor on ground
	Truss roof

Source: NFPA 170, *Standard for Fire Safety Symbols*, NFPA, Quincy, MA, 1999 edition.

STANDARD FIRE SAFETY SYMBOLS

Exhibit VIII.22 Symbols Related to Means of Egress

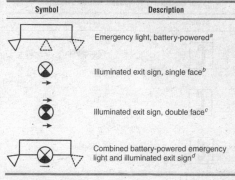

Symbol	Description
	Emergency light, battery-powered[a]
	Illuminated exit sign, single face[b]
	Illuminated exit sign, double face[c]
	Combined battery-powered emergency light and illuminated exit sign[d]

Notes:

[a]Number of lamps on unit to be indicated. Indicate whether light head(s) [lamps(s)] is remote from battery.

[b]Indicate direction of flow for the face.

[c]Indicate direction of flow for each face.

[d]Number of lamps on unit to be indicated. Indicate whether light head(s) [lamp(s)] is remote from battery. Indicate direction of flow for the face.

Source: NFPA 170, Standard for Fire Safety Symbols, NFPA, Quincy, MA, 1999 edition.

STANDARD FIRE SAFETY SYMBOLS

Exhibit VIII.23 Fire Service
or Emergency Telephone Station Symbols

Symbol	Description
ₐ	Accessible
ⱼ	Jack
ₕ	Handset

Source: NFPA 170, *Standard for Fire Safety Symbols,* NFPA, Quincy, MA, 1999 edition.

STANDARD FIRE SAFETY SYMBOLS

Exhibit VIII.24 Suppression System Symbols

Symbol	Description
HL	Halon abort switch
CO_2	Carbon dioxide abort switch
DC	Dry chemical abort switch
FO	Foam abort switch
WC	Wet chemical abort switch
CA	Clean agent abort switch
WM	Water mist abort switch
DL	Deluge sprinkler abort switch

Source: NFPA 170, *Standard for Fire Safety Symbols,* NFPA, Quincy, MA, 1999 edition.

STANDARD FIRE SAFETY SYMBOLS

Exhibit VIII.25 Solenoid Valve Symbol

Source: NFPA 170, *Standard for Fire Safety Symbols*, NFPA, Quincy, MA, 1999 edition.

STANDARD FIRE SAFETY SYMBOLS

Exhibit VIII.26 Smoke/Pressurization Control Symbols

Symbol	Description
	Manual control
	Fan[a]
	Duct
	Roof fan[a]
	Wall fan[a]
	Fire damper
	Smoke damper
	Fire/smoke damper

Continued

STANDARD FIRE SAFETY SYMBOLS

Exhibit VIII.26 Smoke/Pressurization Control Symbols *Continued*

Symbol	Description
	Pressurized stairwell[b]
	Ventilation openings[c]

Notes:

[a]Arrow indicates direction of flow.

[b]Orient as required for base or head injection.

[c]Orient as required for intake or exhaust.

Source: NFPA 170, *Standard for Fire Safety Symbols*, NFPA, Quincy, MA, 1999 edition.

Exhibit VIII.27 Agent Storage Container Symbols

Symbol	Description
FO	Foam
HL	Halon
CO$_2$	Carbon dioxide

Note: Specify type of agent and mounting.

Source: NFPA 170, *Standard for Fire Safety Symbols*, NFPA, Quincy, MA, 1999 edition.

SECTION IX

USEFUL TABLES, FORMULAS, AND FIGURES

This section contains numerous tables that can be used by alarm system designers for making a wide array of conversions as well as other determinations. Conversion factors between SI and U.S. customary units of measurement are provided. Common conversion factors for length, area, volume, light, and temperature are also included. The decimal equivalents included for both inches and feet are very useful when calculating roof pitch.

This section also provides a convenient reference for formulas that are commonly needed in the field. The geometric formulas and reference figures useful for calculating perimeter, area, surface area, and volume of various geometric shapes can be used to begin detector and notification appliance calculations. The trigonometric formulas and reference figures are invaluable to engineers, designers, and installers when determining detector and notification appliance locations in odd-shaped rooms, buildings, and structures.

Exhibit IX.1 Maximum Heat Release Rates—Warehouse Materials

Warehouse Materials	Growth Time (t_g) (s)	Heat Release Density (Q) (Btu/s·ft²)	Classification
1. Wood pallets, stack, 1-1/2 ft high (6%–12% moisture)	150–310	110	fast–medium
2. Wood pallets, stack, 5 ft high (6%–12% moisture)	90–190	330	fast
3. Wood pallets, stack, 10 ft high (6%–12% moisture)	80–110	600	fast
4. Wood pallets, stack, 16 ft high (6%–12% moisture)	75–105	900	fast
5. Mail bags, filled, stored 5 ft high	190	35	medium
6. Cartons, compartmented, stacked 15 ft high	60	200	fast
7. Paper, vertical rolls, stacked 20 ft high	15–28	—	a
8. Cotton (also PE, PE/cot, acrylic/nylon/PE), garments in 12-ft high racks	20–42	—	a
9. Cartons on pallets, rack storage, 15 ft–30 ft high	40–280	—	fast–medium
10. Paper products, densely packed in cartons, rack storage, 20 ft high	470	—	slow
11. PE letter trays, filled, stacked 5 ft high on cart	190	750	medium
12. PE trash barrels in cartons, stacked 15 ft high	55	250	fast
13. FRP shower stalls in cartons, stacked 15 ft high	85	110	fast
14. PE bottles, packed in item 6	85	550	fast
15. PE bottles in cartons, stacked 15 ft high	75	170	fast
16. PE pallets, stacked 3 ft high	130	—	fast
17. PE pallets, stacked 6 ft–8 ft high	30–55	—	fast
18. PU mattress, single, horizontal	110	—	fast

HEAT RELEASE RATE TABLES

19. PE insulation board, rigid foam, stacked 15 ft high	8	170	[a]
20. PS jars, packed in item 6	55	1200	fast
21. PS tubs nested in cartons, stacked 14 ft high	105	450	fast
22. PS toy parts in cartons, stacked 15 ft high	110	180	fast
23. PS insulation board, rigid, stacked 14 ft high	7	290	[a]
24. PVC bottles, packed in item 6	9	300	[a]
25. PP tubs, packed in item 6	10	390	[a]
26. PP and PE film in rolls, stacked 14 ft high	40	350	[a]
27. Distilled spirits in barrels, stacked 20 ft high	23–40	—	[a]
28. Methyl alcohol	—	65	—
29. Gasoline	—	200	—
30. Kerosene	—	200	—
31. Diesel oil	—	180	—

For SI units: 1 ft = 0.305 m.

Notes: The heat release rates per unit floor area are for fully involved combustibles, assuming 100 percent combustion efficiency. The growth times shown are those required to exceed 1000 Btu/s heat release rate for developing fires, assuming 100 percent combustion efficiency. PE = polyethylene, PS = polystyrene, PVC = polyvinyl chloride, PP = polypropylene, PU = polyurethane, and FRP = fiberglass-reinforced polyester.

[a]Fire growth rate exceeds design data.

Source: NFPA 72®, *National Fire Alarm Code*®, NFPA, Quincy, MA, 1999 edition, Table B-2.3.2.3.1(a).

Exhibit IX.2 Characteristics of Ignition Sources

Ignition Source	Typical Heat Output (W)	Burn Time[a] (s)	Maximum Flame Height (mm)	Flame Width (mm)	Maximum Heat Flux (kW/m²)
Cigarette 1.1 g (not puffed, laid on solid surface), bone dry	5	1,200	—	—	42
Conditioned to 50% Relative humidity	5	1,200	—	—	35
Methenamine pill, 0.15 g	45	90	—	—	4
Match, wooden (laid on solid surface)	80	20–30	30	14	18–20
Wood cribs. BS 5852 Part 2					
No. 4 crib, 8.5 g	1,000	190	—	—	15[d]
No. 5 crib, 17 g	1,900	200	—	—	17[d]
No. 6 crib, 60 g	2,600	190	—	—	20[d]
No. 7 crib, 126 g	6,400	350	—	—	25[d]
Crumpled brown lunch bag, 6 g	1,200	80	—	—	—
Crumpled wax paper, 4.5 g (tight)	1,800	25	—	—	—
Crumpled wax paper, 4.5 g (loose)	5,300	20	—	—	—

Folded double-sheet newspaper, 22 g (bottom ignition)	4,000	100	—	—	—
Crumpled double-sheet newspaper, 22 g (top ignition)	7,400	40	—	—	—
Crumpled double-sheet newspaper, 22 g (bottom ignition)	17,000	20	—	—	—
Polyethylene wastebasket, 285 g, filled with 12 milk cartons (390 g)	50,000	200[b]	550	200	35[c]
Plastic trash bags, filled with cellulosic trash (1.2–14 kg)[e]	120,000 to 350,000	200[b]	—	—	—

For SI units, 1 in. = 25.4 mm; 1 Btu/s = 1,055 W; 1 oz = 0.02835 kg = 28.35 g; 1 Btu/ft²-s = 11.35 kW/m².

[a]Time duration of significant flaming.

[b]Total burn time in excess of 1800 seconds.

[c]As measured on simulation burner.

[d]Measured from 25 mm away.

[e]Results vary greatly with packing density.

Source: Merton W. Bunker, Jr., and Wayne D. Moore, eds., *National Fire Alarm Code Handbook*®, NFPA, Quincy, MA, 1999, Table B.5.3(b) and *NFPA 72*®, *National Fire Alarm Code*®, NFPA, Quincy, MA, 1999 edition, Table B-2.3.1(d).

HEAT RELEASE RATE TABLES

Exhibit IX.3 Furniture Heat Release Rates

Test No.	Item/Description/Mass	Growth Time (t_g) (s)	Classification	Fuel Fire Intensity Coefficient (α) (kW/s^2)	Virtual Time (t_v) (s)	Maximum Heat Release Rates (kW)
15	Metal wardrobe, 41.4 kg (total)	50	fast	0.4220	10	750
18	Chair F33 (trial love seat), 29.2 kg	400	slow	0.0066	140	950
19	Chair F21, 28.15 kg (initial)	175	medium	0.0344	110	350
19	Chair F21, 28.15 kg (later)	50	fast	0.4220	190	2000
21	Metal wardrobe, 40.8 kg (total) (initial)	250	medium	0.0169	10	250
21	Metal wardrobe, 40.8 kg (total) (average)	120	fast	0.0733	60	250
21	Metal wardrobe, 40.8 kg (total) (later)	100	fast	0.1055	30	140
22	Chair F24, 28.3 kg	350	medium	0.0086	400	700
23	Chair F23, 31.2 kg	400	slow	0.0066	100	700
24	Chair F22, 31.9 kg	2000	slow	0.0003	150	300
25	Chair F26, 19.2 kg	200	medium	0.0264	90	800
26	Chair F27, 29.0 kg	200	medium	0.0264	360	900
27	Chair F29, 14.0 kg	100	fast	0.1055	70	1850
28	Chair F28, 29.2 kg	425	slow	0.0058	90	700
29	Chair F25, 27.8 kg (later)	60	fast	0.2931	175	700
29	Chair F25, 27.8 kg (initial)	100	fast	0.1055	100	700
30	Chair F30, 25.2 kg	60	fast	0.2931	70	950

31	Chair F31 (love seat), 39.6 kg	60	fast	0.2931	145	2600
37	Chair F31 (love seat), 40.4 kg	80	fast	0.1648	100	2750
38	Chair F32 (sofa), 51.5 kg	100	fast	0.1055	50	3000
39	1/2-in. plywood wardrobe with fabrics, 68.5 kg	35	[e]	0.8612	20	3250
40	1/2-in. plywood wardrobe with fabrics, 68.32 kg	35	[a]	0.8612	40	3500
41	1/8-in. plywood wardrobe with fabrics, 36.0 kg	40	[a]	0.6594	40	6000
42	1/8-in. plywood wardrobe with fire-retardant interior finish	70	fast	0.2153	50	2000
42	1/8-in. plywood wardrobe with fire-retardant interior finish (initial growth) (later growth)	30	[a]	1.1722	100	5000
43	Repeat of 1/2-in. plywood wardrobe, 67.62 kg	30	[a]	1.1722	50	3000
44	1/8-in. plywood wardrobe with fire-retardant latex paint, 37.26 kg	90	fast	0.1302	30	2900
45	Chair F21, 28.34 kg	100	[a]	0.1055	120	2100
46	Chair F21, 28.34 kg	4	[a]	0.5210	130	2600
47	Chair, adj. back metal frame, foam cushions, 20.82 kg	170	medium	0.0365	30	250
48	Eash chair CO7, 11.52 kg	175	medium	0.0344	90	950
49	Easy chair F34, 15.68 kg	200	medium	0.0264	50	200
50	Chair, metal frame, minimum cushion, 16.52 kg	200	medium	0.0264	120	3000
51	Chair, molded fiberglass, no cushion, 5.28 kg	120	fast	0.0733	20	35
52	Molded plastic patient chair, 11.26 kg	275	medium	0.0140	2090	700

Continued

Exhibit IX.3 Furniture Heat Release Rates *Continued*

Test No.	Item/Description/Mass	Growth Time (t_g) (s)	Classification	Fuel Fire Intensity Coefficient (α) (kW/s²)	Virtual Time (t_v) (s)	Maximum Heat Release Rates (kW)
53	Chair, metal frame, padded seat and back, 15.54 kg	350	medium	0.0086	50	280
54	Love seat, metal frame, foam cushions, 27.26 kg	500	slow	0.0042	210	300
56	Chair, wood frame, latex foam cushions, 11.2 kg	500	slow	0.0042	50	85
57	Love seat, wood frame, foam cushions, 54.6 kg	350	medium	0.0086	500	1000
61	Wardrobe, 3/4-in. particleboard, 120.33 kg	150	medium	0.0469	0	1200
62	Bookcase, plywood with aluminum frame, 30.39 kg	65	fast	0.2497	40	25
64	Easy chair, molded flexible urethane frame, 15.98 kg	1000	slow	0.0011	750	450
66	Easy chair, 23.02 kg	76	fast	0.1827	3700	600
67	Mattress and box spring, 62.36 kg (later)	350	medium	0.0086	400	500
67	Mattress and box spring, 62.36 kg (initial)	1100	slow	0.0009	90	400

For SI units: 1 ft = 0.305 m; 1000 Btu/sec = 1055 kW; 1 lb = 0.435 kg.

Note: For tests 19, 21, 29, 42, and 67, different power law curves were used to model the initial and the latter realms of burning. In examples such as these, engineers should choose the fire growth parameter that best describes the realm of burning to which the detection systems is being designed to respond.

[a]Fire growth exceeds design data.

Source: NFPA 72®, National Fire Alarm Code®, NFPA, Quincy, MA, 1999 edition, Table B-2.3.3.1(e).

CONVERSION FACTORS

Exhibit IX.4 Length

1 inch = 0.08333 foot, 1,000 mils, 25.40 millimeters

1 foot = 0.3333 yard, 12 inches, 0.3048 meter, 304.8 millimeters

1 yard = 3 feet, 36 inches, 0.9144 meter

1 rod = 16.5 feet, 5.5 yards, 5.029 meters

1 mile = (U.S. and British) = 5,280 feet, 1.609 kilometers, 0.8684 nautical mile

1 millimeter = 0.03937 inch, 39.37 mils, 0.001 meter, 0.1 centimeter, 100 microns

1 meter = 1.094 yards, 3.281 feet, 39.37 inches, 1,000 millimeters

1 kilometer = 0.6214 mile, 1.094 yards, 3,281 feet, 1,000 meters

1 nautical mile = 1.152 miles (statute), 1.853 kilometers

1 micron = 0.03937 mil, 0.00003937 inch

1 mil = 0.001 inch, 0.0254 millimeters, 25.40 microns

1 degree = 1/360 circumference of a circle, 60 minutes, 3,600 seconds

1 minute = 1/60 degree, 60 seconds

1 second = 1/60 minute, 1/3600 degree

Source: David R. Hague, *NFPA Pocket Guide to Sprinkler System Installation*, NFPA, Quincy, MA, 2001, Table VI.1.

Exhibit IX.5 Area

1 square inch = 0.006944 square foot, 1,273,000 circular mils, 645.2 square millimeters

1 square foot = 0.1111 square yard, 144 square inches, 0.09290 square meter, 92,900 square millimeters

1 square yard = 9 square feet, 1,296 square inches, 0.8361 square meter

1 acre = 43,560 square feet, 4,840 square yards, 0.001563 square mile, 4,047 square meters, 160 square rods

1 square mile = 640 acres, 102,400 square rods, 3,097,600 square yards, 2.590 square kilometers

1 square millimeter = 0.001550 square inch, 1.974 circular mils

Continued

CONVERSION FACTORS

Exhibit IX.5 Area *Continued*

1 square meter = 1.196 square yards, 10.76 square feet, 1,550 square inches, 1,000,000 square millimeters

1 square kilometer = 0.3861 square mile, 247.1 acres, 1,196,000 square yards, 1,000,000 square meters

1 circular mil = 0.7854 square mil, 0.0005067 square millimeter, 0.0000007854 square inch

Source: David R. Hague, *NFPA Pocket Guide to Sprinkler System Installation*, NFPA, Quincy, MA, 2001, Table VI.2.

Exhibit IX.6 Volume (Capacity)

1 fluid ounce = 1.805 cubic inches, 29.57 milliliters, 0.03125 quarts (U.S.) liquid measure

1 cubic inch = 0.5541 fluid ounce, 16.39 milliliters

1 cubic foot = 7.481 gallons (U.S.), 6.229 gallons (British), 1,728 cubic inches, 0.02832 cubic meter, 28.32 liters

1 cubic yard = 27 cubic feet, 46,656 cubic inches, 0.7646 cubic meter, 746.6 liters, 202.2 gallons (U.S.), 168.4 gallons (British)

1 gill = 0.03125 gallon, 0.125 quart, 4 ounces, 7.219 cubic inches, 118.3 milliliters

1 pint = 0.01671 cubic foot, 28.88 cubic inches, 0.125 gallon, 4 gills, 16 fluid ounces, 473.2 milliliters

1 quart = 2 pints, 32 fluid ounces, 0.9464 liter, 946.4 milliliters, 8 gills, 57.75 cubic inches

1 U.S. gallon = 4 quarts, 128 fluid ounces, 231.0 cubic inches, 0.1337 cubic foot, 3.785 liters (cubic decimeters), 3,785 milliliters, 0.8327 Imperial gallon

1 Imperial (British and Canadian) gallon = 1.201 U.S. gallons, 0.1605 cubic foot, 277.3 cubic inches, 4.546 liters (cubic decimeters), 4,546 milliliters

1 U.S. bushel = 2,150 cubic inches, 0.9694 British bushel, 35.24 liters

1 barrel (U.S. liquid) = 31.5 gallons (various industries have special definitions of a barrel)

1 barrel (petroleum) = 42.0 gallons

1 millimeter = 0.03381 fluid ounce, 0.06102 cubic inch, 0.001 liter

1 liter (cubic decimeter) = 0.2642 gallon, 0.03532 cubic foot, 1.057 quarts, 33.81 fluid ounces, 61.03 cubic inches, 1,000 milliliters

CONVERSION FACTORS

Exhibit IX.6 Volume (Capacity) *Continued*

1 cubic meter (kiloliter) = 1.308 cubic yards, 35.32 cubic feet, 264.2 gallons, 1,000 liters

1 cord = 128 cubic feet, 8 feet × 4 feet × 4 feet, 3.625 cubic meters

Source: David R. Hague, *NFPA Pocket Guide to Sprinkler System Installation,* NFPA, Quincy, MA, 2001, Table VI.3.

Exhibit IX.7 Velocity

1 foot per second = 0.6818 mile per hour, 18.29 meters per minute, 0.3048 meters per second

1 mile per hour = 1.467 feet per second, 1.609 kilometers per hour, 26.82 meters per minute, 0.4470 meters per second.

1 kilometer per hour = 0.2778 meter per second, 0.5396 knot per hour, 0.6214 mile per hour, 54.68 feet per minute

1 meter per minute = 0.03728 mile per hour, 0.05468 foot per second, 0.06 kilometer per hour, 16.67 millimeters per second, 3.281 feet per minute

1 knot per hour = 1.152 miles per hour, 1.689 feet per second, 1.853 kilometers per hour

1 revolution per minute = 0.01667 revolution per second, 6 degrees per second

1 revolution per second = 60 revolutions per minute, 360 degrees per second

Source: David R. Hague, *NFPA Pocket Guide to Sprinkler System Installation,* NFPA, Quincy, MA, 2001, Table VI.5.

Exhibit IX.8 Pressure

1 atmosphere = pressure exerted by 760 millimeters of mercury of standard density at 0°C, 14.70 pounds per square inch, 29.92 inches of mercury at 32°F, 33.90 feet of water at 39.2°F, 101.3 kilopascal

1 millimeter of mercury (at 0°C) = 0.001316 atmosphere, 0.01934 pound per square inch, 0.04460 foot of water (4°C or 39.2°F), 0.0193 pound per square inch, 0.1333 kilopascal

Continued

CONVERSION FACTORS

Exhibit IX.8 Pressure *Continued*

1 inch of water (at 39.2°F) = 0.00246 atmosphere, 0.0361 pound per square inch, 0.0736 inch of mercury (at 32°F), 0.2491 kilopascal

1 foot of water (at 39.2°F) = 0.02950 atmosphere, 0.4335 pound per square inch, 0.8827 inch of mercury (at 32°F), 22.42 millimeters of mercury, 2.989 kilopascal

1 inch of mercury (at 32°F) = 0.03342 atmosphere, 0.4912 pound per square inch, 1.133 feet of water, 13.60 inches of water (at 39.2°F), 3.386 kilopascal

1 millibar (1/1000 bar) = 0.02953 inch of mercury. A bar is the pressure exerted by a force of one million dynes on a square centimeter of surface

1 pound per square inch = 0.06805 atmosphere, 2.036 inches of mercury, 2.307 feet of water, 51.72 millimeters of mercury, 27.67 inches of water (at 39.2°F), 144 pounds per square foot, 2,304 ounces per square foot, 6.895 kilopascal

1 pound per square foot = 0.00047 atmosphere, 0.00694 pound per square inch, 0.0160 foot of water, 0.391 millimeter of mercury, 0.04788 kilopascal

Absolute pressure = the sum of the gage pressure and the barometric pressure

1 ton (short) per square foot = 0.9451 atmosphere, 13.89 pounds per square inch, 9,765 kilograms per square meter

Source: David R. Hague, *NFPA Pocket Guide to Sprinkler System Installation,* NFPA, Quincy, MA, 2001, Table VI.9.

Exhibit IX.9 Power

1 horse power = 746 watts, 1.014 metric horse power, 10.69 kilograms-calories per minute, 42.42 British thermal units per minute, 550 pound-feet per second, 33,000 pound-feet per minute

1 kilowatt = 1.341 horse power, 1.360 metric horse power, 14.33 kilogram-calories per minute, 56.90 British thermal units per minute, 1,000 watts

Source: David R. Hague, *NFPA Pocket Guide to Sprinkler System Installation,* NFPA, Quincy, MA, 2001, Table VI.10.

CONVERSION FACTORS

Exhibit IX.10 Heat (Mean Values)

1 British thermal unit = 0.2520 kilogram-calorie, 1,055 joules (absolute)

1 kilogram-calorie = 3.969 British thermal units, 4,187 joules

1 British thermal unit per pound = 0.5556 kilogram-calorie per kilogram, 2.325 joules per gram

1 gram-calorie per gram = 1.8 British thermal units per pound, 4.187 joules per gram

Source: David R. Hague, *NFPA Pocket Guide to Sprinkler System Installation,* NFPA, Quincy, MA, 2001, Table VI.11.

Exhibit IX.11 Electrical

1 volt = potential required to produce current flow of 1 ampere through a resistance or impedance of 1 ohm, or current flow of 2 amperes through resistance of 1/2 ohm, etc.

1 ampere = current flow through a resistance or impedance of 1 ohm produced by a potential of 1 volt, or current flow through a resistance of 100 ohms produced by a potential of 100 volts, etc.

1 milliampere = 0.001 ampere

1 ohm = resistance or impedance through which current of 1 ampere will flow under a potential of 1 volt

1 microhm = 0.000001 ohm

mho = unit of conductance. In a direct current, circuit conductance in mhos is the reciprocal of ohms resistance.

1 watt = power developed by current flow of 1 ampere under potential of 1 volt (DC, or AC with power factor unity). See also Power.

1 joule = 1 watt second. A flow of 1 ampere through a resistance of 1 ohm for 1 second. See also Heat.

1 millijoule = 0.001 joule

Source: David R. Hague, *NFPA Pocket Guide to Sprinkler System Installation,* NFPA, Quincy, MA, 2001, Table VI.12.

CONVERSION FACTORS

Exhibit IX.12 Radiation

1 curie = the emission of 3.70×10^{10} beta particles per second (the particles emitted per second from 1 gram of radium)

1 roentgen = the quantity of X-rays which will produce 2.08×10^{9} ion pairs in 1 cubic centimeter of dry air at $0°C$ and standard atmospheric pressure

Source: David R. Hague, *NFPA Pocket Guide to Sprinkler System Installation,* NFPA, Quincy, MA, 2001, Table VI.13.

Exhibit IX.13 Light

Illuminance *E*		The luminous flux per unit area at any point on a surface exposed to light (illumination)
Lumen	lm	The luminous flux of light equal to one candela intensity radiating equally in all directions
Lux	lx	The International System unit of illumination (1 lux = 1 lumen per square meter)
Footcandle	fc	The intensity of light on a surface (1 footcandle = 1 lumen per square foot) (Originally defined as a standardized candle burning at one foot from a given surface. The "standardized" candle was made of Swiss bee's wax with an Egyptian cotton wick, measured at a "standardized" temperature and pressure. Today, it is a piece of platinum glowing at its freezing point excited by an electric current.)

1 lumen (lm) per square meter = 0.09290304 lumen per square foot, 6.4516×10^{-4} lumen per square inch, 1 lux, 0.09290304 footcandles

1 lumen (lm) per square foot = 10.7639104 lumen per square meter, 0.0069444 lumen per square inch, 10.7639104 lux, 1 footcandle

1 lumen (lm) per square inch = 1,550.0031 candela per square meter, 144 candela per square foot, 144 footcandles

Luminance	*I*	The intensity of light per unit area of its source (source intensity)
Candlepower	cp	Luminous intensity *of a steady burning light* expressed in candela
Candela	cd	The unit of luminous intensity

CONVERSION FACTORS

Exhibit IX.13 Light *Continued*

1 candela (cd) per square meter = 0.09290304 candela per square foot, 6.4516×10^{-4} candela per square inch

1 candela (cd) per square foot = 10.7639104 candela per square meter, 0.0069444 candela per square inch

1 candela (cd) per square inch = 1,550.0031 candela per square meter, 144 candela per square foot

Candela-Second	cd	Visible requirements for fire alarm notification appliances are based on a flashing light rather than a steady burning light. In order to minimize confusion between the steady burning light (expressed in candela) and the flashing light, the candela-second (expressed in candela) term is used.
Candela Effective		The effective intensity of the flashing fire alarm notification appliance.

Illuminance and Luminance

1 candela = 12.57 lumens

$E = I/d^2$

$I = E \times d^2$

Exhibit IX.14 Temperature

Temperature Celsius = 5/9 (temperature Fahrenheit − 32°)

Temperature Fahrenheit = 9/5 × temperature Celsius + 32°

Rankine (Fahrenheit absolute) = temperature Fahrenheit + 459.67°

Kelvin (Celsius absolute) = temperature Celsius + 273.15°

Freezing point of water: Celsius = 0°; Fahrenheit = 32°

Boiling point of water: Celsius = 100°; Fahrenheit = 212°

Absolute zero: Celsius = −273.15°; Fahrenheit = −459.67°

Source: David R. Hague, NFPA Pocket Guide to Sprinkler System Installation, NFPA, Quincy, MA, 2001, Table VI.14.

CONVERSION FACTORS

Exhibit IX.15 Celsius to Fahrenheit Temperatures

Celsius	Fahrenheit	Celsius	Fahrenheit
−273.15	−459.67	15	59
−200	−328	15.6	60
−100	−148	16	60.8
0	32	16.1	61
0.56	33	16.7	62
1	33.8	17	62.6
1.11	34	17.2	63
1.67	35	17.8	64
2	35.6	18	64.4
2.22	36	18.3	65
2.78	37	18.9	66
3	37.4	19	66.2
3.33	38	19.4	67
3.89	39	20	68
4	39.2	20.6	69
4.44	40	21	69.8
5	41	21.1	70
5.56	42	21.7	71
6	42.8	22	71.6
6.11	43	22.2	72
6.67	44	22.8	73
7	44.6	23	73.4
7.22	45	23.3	74
7.78	46	23.9	75
8	46.4	24	75.2
8.33	47	24.4	76
8.89	48	25	77
9	48.2	25.6	78
9.44	49	26	78.8
10	50	26.1	79
10.6	51	27	80.6
11	51.8	27.2	81
11.1	52	27.8	82
11.7	53	28	82.4
12	53.6	28.3	83
12.2	54	28.9	84
12.8	55	29	84.2
13	55.4	29.4	85
13.3	56	30	86
13.9	57	30.6	87
14	57.2	31	87.8
14.4	58	31.1	88

CONVERSION FACTORS

Exhibit IX.15 Celsius to Fahrenheit Temperatures *Continued*

Celsius	Fahrenheit	Celsius	Fahrenheit
31.7	89	60	140
32	89.6	61	141.8
32.2	90	62	143.6
32.8	91	63	145.4
33	91.4	64	147.2
33.3	92	65	149
33.9	93	65.6	150
34	93.2	66	150.8
34.4	94	67	152.6
35	95	68	154.4
35.6	96	69	156.2
36	96.8	70	158
36.1	97	71	159.8
36.7	98	71.1	160
37	98.6	72	161.6
37.2	99	73	163.4
37.8	100	74	165.2
38	100.4	75	167
39	102.2	76	168.8
40	104	76.7	170
41	105.8	77	171.6
42	107.6	78	172.4
43	109.4	79	174.2
43.3	110	80	176
44	111.2	81	177.8
45	113	82	179.6
46	114.8	82.2	180
47	116.6	83	181.4
48	118.4	84	183.2
48.9	120	85	185
49	120.2	86	186.8
50	122	87	188.6
51	123.8	87.8	190
52	125.6	88	190.4
53	127.4	89	192.2
54	129.2	90	194
54.4	130	91	195.8
55	131	92	197.6
56	132.8	93	199.4
57	134.6	93.3	200
58	136.4	94	201.2
59	138.2	95	203

Continued

CONVERSION FACTORS

Exhibit IX.15 Celsius to Fahrenheit Temperatures *Continued*

Celsius	Fahrenheit	Celsius	Fahrenheit
96	204.8	760	1400
97	206.6	800	1472
98	208.4	871.1	1600
98.9	210	900	1652
99	210.2	982.2	1800
100	212	1000	1832
120	248	1093.3	2000
121.1	250	1100	2012
140	284	1200	2192
148.9	300	1204.4	2200
160	320	1300	2372
176.7	350	1315.6	2400
180	356	1400	2552
200	392	1428	2600
204.4	400	1500	2732
250	482	1537.8	2800
260	500	1600	2912
300	572	1648.9	3000
315.8	600	1700	3092
350	662	1760	3200
371.1	700	1800	3272
400	752	1871.1	3400
426.7	800	1900	3452
450	842	1982	3600
482.2	900	2000	3632
500	932	2204.4	4000
537.8	1000	2500	4532
600	1112	2760	5000
648.9	1200	3000	5432
700	1292		

Source: David R. Hague, *NFPA Pocket Guide to Sprinkler System Installation*, NFPA, Quincy, MA, 2001, Table VI.15.

GEOMETRIC FORMULAS

Exhibit IX.16 Perimeter

(a) Square

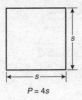

$$P = 4s$$

(d) Triangle

$$P = a + b + c$$

(b) Rectangle

$$P = 2(b + h)$$

(e) Trapezoid

$$P = a + b + c + d$$

(c) Parallelogram

$$P = 2(b + a)$$

(f) Circle

$$C = 2\pi r = \pi d$$

a, b, c, d	= sides	P	= perimeter
C	= circumference	π	= 3.14159
D	= diameter	r	= radius
h	= altitude	s	= side of square

Source: David R. Hague, *NFPA Pocket Guide to Sprinkler System Installation,* NFPA, Quincy, MA, 2001, Figure V.1.

GEOMETRIC FORMULAS

Exhibit IX.17 Area

(a) Square

$$A = s^2$$

(e) Trapezoid

$$A = h\frac{(b + B)}{2}$$

(b) Rectangle

$$A = bh$$

(f) Circle

$$A = \pi r^2 = 0.7854 D^2$$

(c) Parallelogram

$$A = bh$$

(g) Quadrilateral

$$A = \frac{ac \times bd \times \sin \theta}{2}$$

(d) Triangle

$$A = \frac{bh}{2}$$

A	=	area
a, b, c, d	=	sides
B	=	large base
b	=	base or small base
D	=	diameter
h	=	altitude
π	=	3.14159
r	=	radius
s	=	side of square

Source: David R. Hague, *NFPA Pocket Guide to Sprinkler System Installation*, NFPA, Quincy, MA, 2001, Figure V.2.

GEOMETRIC FORMULAS

Exhibit IX.18 Volume

(a) Cube

$$V = e^3$$

(e) Sphere

$$V = \frac{4\pi r^3}{3}$$

(b) Prism

$$V = lwh$$

(f) Cylinder

$$V = \pi r^2 h$$

(c) Pyramid

$$V = \frac{Ah}{3}$$

(g) Cone

$$V = \frac{\pi r^2 h}{3}$$

(d) Roof space

$$V = \frac{h \times c \times b}{2}$$

A	= area	l	= length
b, c	= sides	π	= 3.14159
D	= diameter	r	= radius
e	= edge of cube	V	= volume
h	= height	w	= width

Source: David R. Hague, NFPA Pocket Guide to Sprinkler System Installation, NFPA, Quincy, MA, 2001, Figure V.3.

GEOMETRIC FORMULAS

Exhibit IX.19 Arcs and Chords

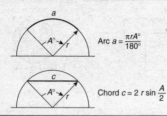

Arc $a = \dfrac{\pi r A^\circ}{180^\circ}$

Chord $c = 2\,r \sin\dfrac{A}{2}$

Source: David R. Hague, *NFPA Pocket Guide to Sprinkler System Installation,* NFPA, Quincy, MA, 2001, Figure V.8.

TRIGONOMETRIC FORMULAS

Exhibit IX.20 Oblique Triangles

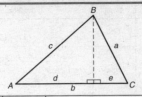

Find	Given	Solution
a	ABb	$b \sin A \div \sin B$
	ABc	$c \sin A \div \sin (A + B)$
	ACb	$b \sin A \div \sin (A + C)$
	ACc	$c \sin A \div \sin C$
	BCb	$b \sin (B + C) \div \sin B$
	BCc	$c \sin (B + C) \div \sin C$
	Abc	$\sqrt{b^2 + c^2 - 2bc \cdot \cos A}$
b	ABa	$a \sin B \div \sin A$
	ABc	$c \sin B \div \sin (A + B)$
	ACa	$a \sin (A + C) \div \sin A$
	ACc	$c \sin A + C \div \sin C$
	BCa	$a \sin B \div \sin (B + C)$
	BCc	$c \sin B \div \sin C$
	Bac	$\sqrt{a^2 + c^2 - 2ac \cdot \cos B}$
c	ABa	$a \sin (A + B) \div \sin A$
	ABb	$b \sin (A + B) \div \sin B$
	ACa	$a \sin C \div \sin A$
	ACb	$b \sin C \div \sin (A + C)$
	BCa	$a \sin C \div \sin (B + C)$
	BCb	$b \sin C \div \sin B$
	Cab	$\sqrt{a^2 + b^2 - 2ab \cdot \cos C}$
1/2(B + C) 1/2(B − C)	Abc	$90° - 1/2A$ $\tan \cdot [(b - c) \tan (90° - 1/2A)] \div (b + c)$
1/2(A + C) 1/2(A − C)	Bac	$90° - 1/2B$ $\tan \cdot [(a - c) \tan (90° - 1/2B)] \div (a + c)$

Continued

Exhibit IX.20 Oblique Triangles *Continued*

Find	Given	Solution
1/2(A + B) 1/2(A − B)	Cab	$90° − 1/2C$ $\tan \cdot [(a − b) \tan(90° − 1/2C)] \div (a + b)$
A	abcs	$\sin 1/2A = \sqrt{(s − b)(s − c) \div bc}$ $\cos 1/2A = \sqrt{s(s − c) \div bc}$ $\tan 1/2A = \sqrt{(s − b)(s − c) − s(s − c)}$
	Bab Bac Cab Cac	$\sin A = a \sin B \div b$ $1/2(A + C) + 1/2(A − C)$ $1/2(A + B) + 1/2(A − B)$ $\sin A = a \sin C \div c$
B	abcs	$\sin 1/2B = \sqrt{(s − a)(s − c) − s(s − a)}$ $\cos 1/2B = \sqrt{s(s − b) \div ac}$ $\tan 1/2B = \sqrt{(s − a)(s − c) − s(s − b)}$
	Aab Abc Cab Cac	$\sin B = b \sin A \div a$ $1/2(B + C) + 1/2(B − C)$ $1/2(A + B) − 1/2(A − B)$ $\sin B = b \sin C \div c$
C	abcs	$\sin 1/2C = \sqrt{(s − a)(s − b) \div ab}$ $\cos 1/2C = \sqrt{s(s − c) \div ab}$ $\tan 1/2C = \sqrt{(s − a)(s − b) − s(s − c)}$
	Aac Abc Bac Bbc	$\sin C = c \sin A \div a$ $1/2(B + C) = 1/2(B − C)$ $1/2(A + C) = 1/2(A − C)$ $\sin C = c \sin B \div b$
Area	abc Cab	$\sqrt{s(s − a)(s − b)(s − c)}$ $1/2ab \sin C$
s	abc	$a + b + c \div 2$
d	abcs	$(b^2 + c^2 − a^2) \div 2b$
e	abcs	$(a^2 + b^2 − c^2) \div 2b$

Source: David R. Hague, *NFPA Pocket Guide to Sprinkler System Installation*, NFPA, Quincy, MA, 2001, Table V.1.

TRIGONOMETRIC FORMULAS

Exhibit IX.21 Right Triangles

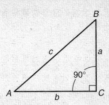

Find	Given	Solution
A	ab	tan A = a ÷ b
	ac	sin A = a ÷ c
	bc	cos A = b ÷ c
B	ab	tan B = b ÷ a
	ac	cos B = a ÷ c
	bc	sin B = b ÷ c
a	Ab	b tan A
	Ac	c sin A
b	Aa	a ÷ tan A
	Ac	c cos A
c	Aa	a ÷ sin A
	Ab	b ÷ cos A
Area	ab	ab ÷ 2

Source: David R. Hague, *NFPA Pocket Guide to Sprinkler System Installation,* NFPA, Quincy, MA, 2001, Table V.2.

Exhibit IX.22 Trigonometric Functions

Angle	Sine	Cosine	Tangent	Cotangent	Secant	Cosecant	Angle
0	.0000	1.0000	.0000		1.0000		90
1	.0175	.9998	.0175	57.2900	1.0002	57.2987	89
2	.0349	.9994	.0349	28.6363	1.0006	28.6537	88
3	.0523	.9986	.0524	19.0811	1.0014	19.1073	87
4	.0698	.9976	.0699	14.3007	1.0024	14.3356	86
5	.0872	.9962	.0875	11.4301	1.0038	11.4737	85
6	.1045	.9945	.1051	9.5144	1.0055	9.5668	84
7	.1219	.9925	.1228	8.1443	1.0075	8.2055	83
8	.1392	.9903	.1405	7.1154	1.0098	7.1853	82
9	.1564	.9877	.1584	6.3138	1.0125	6.3925	81
10	.1736	.9848	.1763	5.6713	1.0154	5.7588	80
11	.1908	.9816	.1944	5.1446	1.0187	5.2408	79
12	.2079	.9781	.2126	4.7046	1.0223	4.8097	78
13	.2250	.9744	.2309	4.3315	1.0263	4.4454	77
14	.2419	.9703	.2493	4.0108	1.0306	4.1336	76
15	.2588	.9659	.2679	3.7321	1.0353	3.8637	75
16	.2756	.9613	.2867	3.4874	1.0403	3.6280	74
17	.2924	.9563	.3057	3.2709	1.0457	3.4203	73

TRIGONOMETRIC FORMULAS

18	.3090	.9511	.3249	3.0777	1.0515	3.2361	72
19	.3256	.9455	.3443	2.9042	1.0576	3.0716	71
20	.3420	.9397	.3640	2.7475	1.0642	2.9238	70
21	.3584	.9336	.3839	2.6051	1.0711	2.7904	69
22	.3746	.9272	.4040	2.4751	1.0785	2.6695	68
23	.3907	.9205	.4245	2.3559	1.0864	2.5593	67
24	.4067	.9135	.4452	2.2460	1.0946	2.4586	66
25	.4226	.9063	.4663	2.1445	1.1034	2.3662	65
26	.4384	.8988	.4877	2.0503	1.1126	2.2812	64
27	.4540	.8910	.5095	1.9626	1.1223	2.2027	63
28	.4695	.8829	.5317	1.8807	1.1326	2.1301	62
29	.4848	.8746	.5543	1.8040	1.1434	2.0627	61
30	.5000	.8660	.5774	1.7321	1.1547	2.0000	60
31	.5150	.8572	.6009	1.6643	1.1666	1.9416	59
32	.5299	.8480	.6249	1.6003	1.1792	1.8871	58
33	.5446	.8387	.6494	1.5399	1.1924	1.8361	57
34	.5592	.8290	.6745	1.4826	1.2062	1.7883	56
35	.5736	.8192	.7002	1.4281	1.2208	1.7434	55
36	.5878	.8090	.7265	1.3764	1.2361	1.7031	54
37	.6018	.7986	.7536	1.3270	1.2521	1.6616	53
38	.6157	.7880	.7813	1.2799	1.2690	1.6243	52

Continued

TRIGONOMETRIC FORMULAS

Exhibit IX.22 Trigonometric Functions *Continued*

Angle	Sine	Cosine	Tangent	Cotangent	Secant	Cosecant	Angle
39	.6293	.7771	.8098	1.2349	1.2868	1.5890	51
40	.6428	.7660	.8391	1.1918	1.3054	1.5557	50
41	.6561	.7547	.8693	1.1504	1.3250	1.5243	49
42	.6691	.7431	.9004	1.1106	1.3456	1.4945	48
43	.6820	.7314	.9325	1.0724	1.3673	1.4663	47
44	.6947	.7193	.9657	1.0355	1.3902	1.4396	46
45	.7071	.7071	1.0000	1.0000	1.4142	1.4142	45

MISCELLANEOUS DATA

Exhibit IX.23 Prefixes

Prefix	Symbol	Multiplying Factor	
exa-	E	10^{18}	= 1,000,000,000,000,000,000
peta-	P	10^{15}	= 1,000,000,000,000,000
tera-	T	10^{12}	= 1,000,000,000,000
giga-	G	10^{9}	= 1,000,000,000
mega-	M	10^{6}	= 1,000,000
kilo-	K	10^{3}	= 1,000
hecto-	h	10^{2}	= 100
deca-	da	10	= 10
deci-	d	10^{-1}	= 0.1
centi-	c	10^{-2}	= 0.01
milli-	m	10^{-3}	= 0.001
micro-	μ	10^{-6}	= 0.000,001
nano-	n	10^{-9}	= 0.000,000,001
pico-	p	10^{-12}	= 0.000,000,000,001
femto-	f	10^{-15}	= 0.000,000,000,000,001
atto-	a	10^{-18}	= 0.000,000,000,000,000,001

Exhibit IX.24 Electrical Conductivity of Common Materials

Materials	Chemical Symbol	Electrical Conductivity (% of Copper)
Aluminum	AL	64.90
Antimony	SB	4.42
Arsenic	AS	4.90
Beryllium	BE	9.32
Bismuth	Bi	1.50
Brass (70-30)		28.00
Bronze (5% SN)		18.00
Cadmium	CD	22.70
Calcium	CA	50.10
Cobalt	CO	17.80
Copper	CU	
Rolled		100.00
Tubing		100.00
Gold	AU	71.20
Graphite		10^3
Indium	IN	20.60
Iridium	IR	32.50
Iron	FE	17.60
Malleable		10.00
Wrought		10.00
Lead	PB	8.35
Magnesium	MG	38.70
Manganese	MN	0.90
Mercury	HG	1.80
Molybdenum	MO	36.10
Monel (63-37)		3.00
Nickel	NI	25.00
Phosphorous	P	10^{-17}
Platinum	PT	17.50
Potassium	K	28.00
Selenium	SE	14.40
Silicon	SI	10^{-5}
Silver	AG	106.00
Steel (Carbon)		10.00
Stainless		
(18-8)		2.50
(13-CR)		3.50

MISCELLANEOUS DATA

Exhibit IX.24 Electrical Conductivity of Common Materials *Continued*

Materials	Chemical Symbol	Electrical Conductivity (% of Copper)
Tantalum	TA	13.90
Tellurium	TE	10^{-5}
Thorium	TH	9.10
Tin	SN	15.00
Titanium	TI	2.10
Tungsten	W	31.50
Uranium	U	2.80
Vanadium	V	6.63
Zinc	ZN	29.10
Zirconium	ZR	4.20

Exhibit IX.25 Mean Annual Days with Thunderstorms

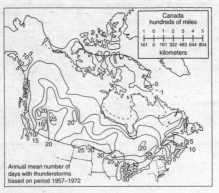

Annual mean number of days with thunderstorms based on period 1957–1972

Source: Richard W. Bukowski and Robert J. O'Laughlin, *Fire Alarm Signaling Systems,* 2nd ed., NFPA, Quincy, MA; Society of Fire Protection Engineers, Boston, 1994, Figure 13-23.

SECTION X

USEFUL CONTACTS

This section provides a listing of organizations and contact information useful to designers, installers, and authorities having jurisdiction.

LIST OF RESOURCES

ADA
Office on the Americans with Disabilities Act
Civil Rights Division
U.S. Division of Justice
P.O. Box 66118
Washington, DC 20035-6118
202-514-0301

AFAA
Automatic Fire Alarm Association, Inc.
P.O. Box 951807
Lake Mary, FL 32795-1807
407-322-6288
407-322-7488 (fax)
e-mail: fire-alarm@afaa.org
www.afaa.org

ANSI
American National Standards Institute
1819 L Street, NW, 6th Floor
Washington, DC, 20036
202-293-8020
202-293-9287 (fax)
e-mail: info@ansi.org
www.ansi.org

ASHRAE
American Society of Heating, Refrigerating and
Air-Conditioning Engineers, Inc.
1791 Tullie Circle N.E.
Atlanta, GA 30329
800-527-4723
404-321-5478 (fax)
www.ashrae.org

ASME   American Society of Mechanical Engineers
             International
             ASME International
             Three Park Avenue
             New York, NY 10016-5990
             800-THE-ASME 800-843-2763
             973-882-1717 (fax)
             email: infocentral@asme.org
             www.asme.org

ASTM   American Society for Testing and Materials
             100 Bar Harbor Drive
             West Conshohocken, PA 19428-2959
             610-832-9500
             610-832-9555 (fax)
             e-mail: webmastr@astm.org
             www.astm.org

ATBCB   Architectural & Transportation Barriers
             Compliance Board
             111 18th Street N.W.
             Suite 501
             Washington, DC 20036-3894
             800-USA-ABLE

CSA   CSA International
             (Canadian Standards Association)
             178 Rexdale Boulevard
             Etobicoke, Ontario, Canada
             M9W 1R3
             800-463-6727
             e-mail: emc@csa-international.org
             www.csa-international.org

CSAA   Central Station Alarm Association
             440 Maple Avenue East
             Suite 201
             Vienna, VA 22180
             703-242-4670
             703-242-4675 (fax)
             e-mail: communications@csaaul.org
             www.csaaul.org

LIST OF RESOURCES

CPSC   U.S. Consumer Product Safety Commission
4330 East-West Highway
Bethesda, MD 20814-4408
800-638-2772 Hot Line
301-504-0990
301-504-0124 (fax)
301-504-0025 (fax)
e-mail: info@cpsc.gov

FDI   Fire Detection Institute
689 Park Avenue
Cincinnati, OH 45246
513-742-4743
513-742-9343 (fax)
e-mail: jmoore@haifire.com
www.firedetinst.org

FM   FMGlobal
1301 Atwood Avenue
Johnston, RI 02919
401-275-3000
401-275-3029 (fax)
e-mail: information@fmglobal.com
www.fmglobal.com

IAEI   International Association
    of Electrical Inspectors
901 Waterfall Way
Suite 602
Richardson, TX 75080-7702
800-786-4234
972-235-6858 (fax)
www.iaei.org

IAFC   International Association of Fire Chiefs
4025 Fair Ridge Drive
Suite 300
Fairfax, VA 22033-2868
703-273-0911
703-273-9363 (fax)
e-mail: dircomm@ichiefs.org
www.ichiefs.org

LIST OF RESOURCES

IBEW   International Brotherhood of Electrical Workers
1125 15th Street, N.W.
Washington, DC 20005
202-833-7000
202-728-7664 (fax)
e-mail: web@ibew.org
www.ibew.org

ICEA   Insulated Cable Engineers Association, Inc.
P.O. Box 1568
Carrollton, GA 30117
770-830-0369
770-830-8501 (fax)
e-mail: support@icea.net
www.icea.net

IEEE   Institute of Electrical and Electronics
         Engineers, Inc.
445 Hoes Lane
P.O. Box 1331
Piscataway, NJ 08855-1331
800-678-IEEE 800-678-4333
732-981-9667 (fax)
e-mail: webmaster@ieee.org
www.ieee.org

IES   Illuminating Engineering Society
         of North America
120 Wall Street, 17th Floor
New York, NY 10005
212-248-5000
212-248-5017 (fax)
212-248-5018 (fax)
e-mail: iesna@iesna.org
www.iesna.org

LIST OF RESOURCES

IMSA   IMSA
          P.O. Box 539
          165 East Union Street
          Newark, NY 14513-0539
          800-723-IMSA 800-723-4672
          315-331-8205 (fax)
          e-mail: info@imsasafety.org
          www.imsasafety.org

ITS   Intertek Testing Services
          ETL SEMCO
          3933 US Route 11
          Industrial Park
          Cortland, NY 13045
          607-753-6711
          607-756-9891 (fax)
          e-mail: info@etlsemko
          www.etlsemko.com

LPI   Lightning Protection Institute
          3335 North Arlington Heights
          Arlington Road
          Suite E
          Arlington Heights, IL 60004
          800-488-6864
          847-577-7276 (fax)
          e-mail: strike@lightning.org
          www.lightning.org

NAFED   National Association of Fire Equipment
              Distributors
          401 North Wabash Avenue
          Suite 732
          Chicago, IL 60611
          312-245-9300
          312-245-9301 (fax)
          e-mail: nafed@nafed.org
          www.nafed.org

LIST OF RESOURCES

NBFAA National Burglar & Fire Alarm Association
8300 Colesville Road
Suite 750
Silver Spring, MD 20910
301-585-1855
301-585-1866 (fax)
e-mail: staff@alarm.org
www.alarm.org

NECA National Electrical Contractors Association
3 Bethesda Metro Center
Suite 1100
Bethesda, MD 20814
301-657-3110
301-215-4500 (fax)
e-mail: webmaster@necanet.org
www.necanet.org

NEMA National Electrical Manufacturers Association
1300 North 17th Street
Suite 1847
Rosslyn, VA 22209
703-841-3200
703-841-5900 (fax)
e-mail: webmaster@nema.org
www.nema.org

NETA International Electrical Testing Association
P.O. Box 687
Morrison, CO 80465-0687
303-697-8441
303-697-8431 (fax)
e-mail: neta@netaworld.org
www.netaworld.org

LIST OF RESOURCES

NJATC   National Joint Apprenticeship
and Training Committee
301 Prince George's Boulevard
Suite D
Upper Marlboro, MD 20774
301-715-2300
301-715-2323 (fax)
e-mail: office@njatc.org
www.njatc.org

NCSBCS   National Conference of States on Building
Codes and Standards
505 Huntmar Park Drive
Suite 210
Herndon, VA 20170-5139
703-437-0100
703-481-3596 (fax)
www.ncsbsc.org

NICET   National Institute for Certification in
Engineering Technologies
1420 King Street
Alexandria, VA 22314-2794
888-IS-NICET 888-476-4238
www.nicet.org

NIST   National Institute of Standards and Technology
Building and Fire Research Laboratory
100 Bureau Drive
Stop 8600
Gaithersburg, MD 20899-8600
301-975-NIST
301-975-6478
301-975-4052 (fax)
e-mail: webmaster@nist.gov
www.nist.gov

LIST OF RESOURCES

OSHA   Occupational Safety & Health Association
U.S. Department of Labor
Office of Public Affairs
Room N3647
200 Constitution Avenue, N.W.
Washington, DC 20210
800-321-OSHA
800-321-6742
www.osha.gov

SFPE   Society of Fire Protection Engineers, Inc.
7315 Wisconsin Avenue
Suite 1225 W
Bethesda, MD 20814
301-718-2910
301-718-2242 (fax)
e-mail: president@sfpe.org
www.sfpe.org

UL   Underwriters Laboratories, Inc.
333 Pfingsten Road
Northbrook, IL 60062-2096
847-272-8800
847-272-8129 (fax)
e-mail: northbrook@us.ul.com
www.ul.com

INDEX